Theodor Reye, Thomas F. Holgate

Lectures on the geometry of position

Theodor Reye, Thomas F. Holgate

Lectures on the geometry of position

ISBN/EAN: 9783742892515

Manufactured in Europe, USA, Canada, Australia, Japa

Cover: Foto ©berggeist007 / pixelio.de

Manufactured and distributed by brebook publishing software
(www.brebook.com)

Theodor Reye, Thomas F. Holgate

Lectures on the geometry of position

LECTURES ON THE
GEOMETRY OF POSITION

BY

THEODOR REYE

PROFESSOR OF MATHEMATICS IN THE UNIVERSITY OF STRASSBURG

TRANSLATED AND EDITED BY

THOMAS F. HOLGATE, M.A., Ph.D.

PROFESSOR OF APPLIED MATHEMATICS IN NORTHWESTERN UNIVERSITY

PART I.

New York

THE MACMILLAN COMPANY

LONDON: MACMILLAN AND CO., LIMITED

1898

All rights reserved

TRANSLATOR'S PREFACE.

In preparing this translation of Professor Reye's *Geometrie der Lage*, my sole object has been to place within easy reach of the English-speaking student of pure geometry an elementary and systematic development of modern ideas and methods. The increasing interest in this study during recent years has seemed to demand a text-book at once scientific and sufficiently comprehensive to give the student a fair view of the field of modern pure geometry, and also sufficiently suggestive to incite him to investigation. The recognized merit of Professor Reye's work in all these regards is my only apology for offering this translation as an attempt to satisfy our present needs.

It has been my aim to present in fair readable English the geometric ideas contained in the text, rather than to hold myself, at all points, to a literal translation; yet I trust that I have not altogether destroyed the charm of the original writing. Some changes have been made; the articles have been numbered, the examples set at the end of the lectures to which they are related and a few new ones added, explanatory notes have been inserted where they seemed necessary or helpful, and an index has been compiled. I have not deemed it advisable to omit from this edition any part of the original prefaces or introduction, even though, at this distance from their first publication, they might not be demanded in their entirety.

For the most part I have endeavoured to hold rigorously to well-established terminology. A few instances of deviation from this principle, however, may be mentioned. I have preferred the terms 'sheaf of rays,' 'sheaf of planes,' and 'bundle of rays or planes,' to the more common though I think less expressive terms, 'flat pencil,' 'axial pencil,' and 'sheaf of lines or planes'; instead of the expression 'conformal representation' as an equivalent for the German '*conforme*

Abbildung,' I have ventured 'conformal depiction.' The term 'ideal' has elsewhere been applied to infinitely distant points and lines; with this I have associated the word 'actual' to apply to points and lines of the finite region.

I desire to acknowledge my indebtedness to my colleague Professor Henry S. White for valuable assistance; my thanks also and the gratitude of all who may profit by the use of this translation are due to Dr. M. C. Bragdon of Evanston whose interest and generosity made its publication possible.

What is commonly known as Modern Synthetic Geometry has been developed for the most part during the present century. It differs from the geometry of earlier times, not so much by the subjects dealt with and the theorems propounded, as by the processes which are employed and the generality of the results which are attained.

Geometry was to the ancients a subject of entrancing interest. Its progress is prominently connected with the names of Thales of Miletus (640-546), Pythagoras (569-500), Plato (429-348), who cultivated geometry as fundamental to the study of philosophy, Menaechmus (375-325), the first to discuss the conic sections, Euclid of Alexandria (330-275), Archimedes (287-212), and Apollonius of Perga (260-200); these among others before the Christian era.

Of the numerous writings of Euclid, the *Elements*,[1] in which was collected and systematized much of the geometrical knowledge of that time, has remained for two thousand years a marvellous monument to his skill. Whatever may be its defects, and these have been the subject of much discussion, it "certainly possesses some excellent features; it accustoms the mind to rigor, to elegance of demonstration, and to the methodical arrangement of ideas; in these regards it is worthy of our admiration."[2] His *Porisms*, which unfortunately have been lost, are said to have contained many of the principles that have formed the basis of modern geometry.

Ancient geometry reached its highest perfection under Archimedes and Apollonius, the former of whom devoted much study to physical problems by means of geometry, and the latter carried his investi-

---

[1] For a convenient summary and characterization of Euclid's *Elements*, see Professor Henrici's article on Geometry in the *Encyclopedia Britannica*.

[2] Poncelet, *Propriétés Projectives*, etc., p. 15.

gations upon the conic sections so far as to leave few of their important properties undiscovered. He produced a systematic treatise on conic sections containing his own discoveries, and including also all previous knowledge of these curves.[1]

The great geometer and commentator of the early centuries of the Christian era was Pappus of Alexandria. In his *Mathematicae Collectiones*, written toward the end of the fourth century, he collated the scattered works of the earlier celebrated geometers and a multitude of curious theorems from many sources, to which he added so much of original work as to place him among the most illustrious of ancient geometers. This work is the chief source of information on ancient geometry. It comments so fully upon Euclid's book of porisms that frequent efforts have been made to restore the latter, notably by Chasles in 1860.

The work of the ancient geometers was fragmentary. Truly remarkable discoveries were made, but general principles were not brought into prominence; theorems were announced disconnectedly as though they had been received by their authors ready made; the method of their discovery was rarely, if ever, indicated; the demonstrations were given in the most polished and systematic form, but the relations existing among different theorems were not shown, and no suggestions were offered for further investigation; special cases of general theorems were as a rule treated as though they were separate and independent theorems.

But, scattered here and there, throughout the great volume of geometrical knowledge accumulated by these early geometers, is to be found the material upon which the beautiful and symmetrical structure of modern geometry has been founded. For example, the property of perspective triangles of which use is made in the geometrical definition of harmonic points, though usually credited to Desargues, was in fact announced by Euclid.[2] Harmonic division itself was known to Apollonius, and the fact that the anharmonic ratio of four collinear points is unaltered by projection was demonstrated by Pappus,[3] and was probably known much earlier. The theorem upon which Carnot based the theory of transversals was discovered and published by Menelaus early in the second century.

---

[1] An edition of Apollonius' *Conic Sections*, with notes, etc., by T. L. Heath, M.A., has recently been published by Macmillan & Co., London.

[2] Pappus, *Mathematicae Collectiones*, preface to book VII.

[3] *Mathematicae Collectiones*, VII., 129.

As has already been suggested, modern geometry is characterized by generality both in its processes and in its results. The founding of modern pure geometry is usually accredited to Monge (1746-1818), whose lectures in the Polytechnic School at Paris were published under the title of *Géométrie Descriptive.* These lectures, by utilizing the theory of transversals and the principle of parallel projection, called attention to the advantages to be gained through the application of geometrical methods, and served to revive the interest in pure geometry, which had been dormant for so many years.

But the generalizing processes which characterize modern geometry were begun much earlier than the time of Monge. Desargues (1593-1662), a contemporary of Descartes, introduced the notion of infinitely distant points and lines, with its far reaching results, and announced the doctrine of continuity. The methods of Pascal (1623-1662) too, so far as it is possible to judge from the few remaining fragments of his mathematical work, partook of the broadest generality, and it is fair to assume that had not the work of these two great geometers been almost entirely lost, and had not their ideas been wholly pushed aside through the overwhelming influence of Descartes' discoveries, many of the geometrical theories and results of the present century would have been developed long ago, and the so-called modern geometry would have been of much earlier date.

As it was, however, pure geometry was but little cultivated for over a hundred years before the time of Monge. Geometrical knowledge was truly increased during this period, especially by Newton (1642-1727), Maclaurin (1698-1746), Robert Simson (1687-1768), and Matthew Stewart (1717-1785), but their methods could scarcely be said to partake of the spirit of modern geometry, and differed but little, if at all, from the methods of the ancient geometers.

The illustrious names in connection with the development of modern pure geometery are Poncelet (1788-1867), Steiner (1796-1863), Von Staudt (1798-1867), and Chasles (1793-1880); and if it were permissible to add the names of living men I should mention Cremona and Reye.

Poncelet's great work, *Traité des propriétés projectives des figures,* etc., appeared in 1822, and at once clearly justifies any claim that may be set up in his behalf as the leader in the so-called modern methods. In this work the principle of continuity, the principle of reciprocity or duality, and the method of projection are the chief factors.

There has been much discussion from time to time upon the question of priority in the establishment of the principle of duality. Poncelet used the method of polar reciprocity with respect to a conic, and thus derived the dual of any geometrical figure, but it is claimed for Gergonne that it was he who first established duality as an independent principle. The name 'duality' is clearly due to Gergonne.[1]

The principle of continuity, which was first assumed by Kepler and later by Desargues, demands of the geometer as well as the analyst the consideration of imaginary quantities. Monge discovered that the results obtained from a geometrical construction would not be invalidated if in a different phase of the construction certain of the points and lines disappeared. Poncelet devoted much attention to imaginary solutions of geometrical problems, but it remained for Von Staudt to build up and to bring to a fair degree of perfection a general theory of geometrical imaginaries.

A conception of the geometer's notion of imaginary quantities can probably be best obtained from the following quotation from Professor H. J. S. Smith:[2]

"All attempts to construct imaginaries have been wholly abandoned in pure geometry; but, by asserting once for all the principle of continuity as universally applicable to all the properties of figured space, geometers have succeeded, if not in explaining the nature of imaginaries, yet, at least, in deriving from them great advantages. They consider it a consequence of the law of continuity that if we once demonstrate a property for any figure in any one of its general states, and if we then suppose the figure to change its form, subject of course to the conditions in accordance with which it was first traced, the property we have proved, though it may become un-meaning, can never become untrue, even if every point and every line by means of which it was originally proved should disappear."

The line of demarcation which was visible as early as the time of Archimedes and Apollonius between the geometers whose theories rest upon metric properties and those whose basal notions are purely positional was very prominent during the foundation period in the development of modern pure geometry. Steiner and Chasles based their investigations upon metric properties, defining the pro-

---

[1] *Annales de Mathématiques,* T. XVI., 1826.
[2] *Collected Papers,* Vol. I. p. 4.

jective relation by means of the anharmonic ratio; Von Staudt, on the other hand, and after him Reye, define this relation by reference to harmonic division, and this in turn is defined purely geometrically. Upon such a definition of projectivity they have been able to perfect a complete theory without any reference to metric properties whatsoever. Cremona avoids metric properties in his foundations by defining two projective forms as the first and last of a series of forms in perspective.

In all the recent development of synthetic geometry the effect of contact with analysis is clearly seen. Through its influence the foundations upon which the science rests have been carefully examined, while characteristic methods of investigation have been acquired. The tendency toward generalization may likewise be attributed largely to the influence of analysis, though it is true that some progress had been made in this direction before the analytical methods had attained such universal sway. But, on the other hand, geometry has done much to enliven and heighten the interest in analysis, so that it may fairly be said that neither pure geometry nor pure analysis can any longer boast an isolated existence, or hope to attain its highest development independently of the other.

THOMAS F. HOLGATE.

EVANSTON, ILLINOIS,
December 1897.

A NOTE FROM THE AUTHOR.

DEAR SIR,—It is with great pleasure and satisfaction that I greet your English translation of my *Geometrie der Lage*, which henceforth will take its place along with the French and Italian translations. I trust that it may help to win for pure Geometry many friends and investigators in the broad English-speaking countries.—I am, yours faithfully,

TH. REYE.

STRASSBURG, *September* 1896.

THE AUTHOR'S PREFACE TO THE FIRST
EDITION OF PART I.

THE Lectures upon the Geometry of Position, which I now offer to the public, have been written at intervals during the last two years. I have been induced to publish them by a need which has been felt for a long time in the technical schools of this country, and perhaps in wider circles. The important graphical methods with which Professor Culmann has enriched the science of engineering, and which are published in his work, *Die graphische Statik*, are based for the most part upon modern geometry, and a knowledge of this subject has therefore become indispensable to students of engineering in our institutions. In the present work I attempt to supply the want of a text-book which offers to the student the necessary material in concise form, and which will be of assistance to me in my oral instruction.

I was obliged, as a matter of course, to make use of the terminology adopted by Culmann, and, to a certain extent, to follow the subject matter of that complete work bearing the same title as this, namely,—*The Geometry of Position*, by Professor Von Staudt. The new terms which Von Staudt added to the older ones of Steiner are so happily chosen that I have preferred a different one in but a single instance, the term "range of points" (*Punktreihe*), first used by Paulus (and Göpel) instead of "line form" (*gerades Gebilde*). The way in which Von Staudt establishes this science in contrast with all other writers upon Modern Geometry appears to me to afford advantages so important that, laying aside all other considerations, I should prefer it to every other. Permit me, in a few words, to assign my reasons for this preference.

To the engineer as well as to the mechanic and architect, the ability to form beforehand a mental picture of his structure as it will appear in space is of great service in designing it. Suppose, for example, a bridge is to be built across a stream. From among the different possible modes of construction, that one must be chosen at the outset which is best adapted to the given conditions. To this end the engineer compares the long iron girders with the boldly swung arch or the freely hanging suspension bridge, and endeavours to conceive how the pressure would be exerted at this point and at that point, and how distributed among the different members of the huge structure. Again and again he examines and compares, goes more and more deeply into all the details, until the whole structure stands complete in clear outline before his mind's eye. And now the second part of his creative work begins. The project is transferred to paper; all details as to form and strength are completely determined. But still the engineer, and everybody else who wishes to become familiar with his ideas, must continually exert his power of imagination in order actually to see the object intended to be represented by the lines of a drawing which is not at all intelligible to the uninitiated. So also the mathematician and in fact any one who concerns himself with the natural sciences must, like the technologist, bring the imagination very frequently into play. At one time he tries to understand a complicated piece of apparatus from an insufficient sketch; at another, from a scanty description, to make intelligible remotely connected processes of nature or complicated motions.

One principal object of geometrical study appears to me to be the exercise and the development of the power of imagination in the student, and I believe that this object is best attained in the way in which Von Staudt proceeds. That is to say, Von Staudt excludes all calculations whether more or less complicated which make no demands upon the imagination, and to whose comprehension there is requisite only a certain mechanical skill having little to do with geometry in itself; and instead, arrives at the knowledge of the geometric truths upon which he bases the Geometry of Position by direct visualization. It cannot be denied that this method, like every other, presents its peculiar difficulties; and, what is more, Von Staudt's own work, evidently not written for a beginner, embodies peculiarities which are praiseworthy enough in themselves, but which essentially increase the difficulties of the study. It is especially marked by a scantiness of expression, and a very condensed, almost laconic, form of statement; nothing is said except what is absolutely necessary, rarely is there a word of explanation given, and it is left to the student to form for himself suitable examples illustrative of the theorems, which are enunciated in their most general form. The material, however, is very clearly and systematically arranged; for example, the subjects of projectivity, of the collinear and the reciprocal relation, and of forms in involution, are completely treated before the theories of conic sections and surfaces of the second order are introduced, and Von Staudt thus gains the advantage of being able to prove the properties of forms of the second order all at once; but, on the other hand, the presentation is so abstract that ordinarily the energies of a beginner are quickly exhausted by his study. These features, which, unfortunately, appear to have stood in the way of the well-merited circulation and the general recognition of Von Staudt's work, stamp it as a treatise on Modern Geometry of superior merit, to which we may very appropriately refer, as did the ancient geometers to Euclid. In my lectures written for beginners, however, I must avoid such peculiarities in order not to become unintelligible.

There is one other difficulty inherent in the course itself which I have purposely not avoided, since it must sooner or later be overcome by everybody who desires to comprehend the properties of three-dimensional figures. I refer to the difficulty already mentioned of getting a mental picture of such figures in space, a difficulty with which the beginner has to struggle in the study of descriptive geometry and analytical geometry of space, the surmounting of which, as I have already remarked, I hold to be one of the principal objects of geometrical instruction. In order to make the accomplishment of this end easier for the student, I have added plates of diagrams to my lectures. Von Staudt did not make use of such expedients; in fact, we should not be far from the truth in ascribing to him views similar to those expressed at one time by Steiner, "that stereometric ideas can be correctly comprehended only when they are contemplated purely by the inner power of imagination, without any means of illustration whatever." By disdaining to make use of these instruments of illustration, which so far as planimetric ideas are concerned, are not at all likely to lead to an incorrect conception, I should unnecessarily have increased the difficulty, on the part of the student, of comprehending my lectures.

Since the method introduced by Von Staudt excludes numerical computations, and investigates the metric properties of geometrical figures apart from the general theory, it presents still another advantage to which I should like to call especial attention. That is, it turns to account most beautifully, in all its clearness and to its full compass, the important and fruitful principle of duality or reciprocity, by which the whole Geometry of Position is controlled. No method making use of the idea of measurement can boast of this merit, and for the simple reason that in metric geometry this principle is not in general applicable. But it must be admitted that geometry offers nothing which is so stimulating to the beginner, and which so spurs him on to independent research as the principle of reciprocity, and the earlier he is made acquainted with it the better. The fact that this principle stands out so clearly, particularly in geometry of three dimensions, was for me a determinative reason for not separating the stereometric discussions from the planimetric.

Metric relations, I must add, especially those of the conic sections, have by no means been neglected in my lectures; on the contrary, I have throughout developed these relations to a greater extent than did either Steiner or Von Staudt, wherever they could naturally present themselves as special cases of general theorems.

I proceed to the study of the conic sections and other forms of the second order treated in this first part of my lectures, by a route different from that of Steiner or Von Staudt, the latter of whom based the theory of these forms upon the doctrine of collineation and reciprocity. By introducing the forms of the second order directly from a study of projective one-dimensional primitive forms, I hope to have made the comprehension of the projective relation easier; at the same time, I secure the advantage of being able to prepare the beginner by degrees for the more difficult study of collineation and reciprocity. In his highly suggestive pioneer work, *Systematische Entwickelung*, etc., Steiner has furnished us with the model which I have preferred to follow in my lectures from the fifth to the tenth. For reasons already referred to, however, I was obliged to refrain from defining conic sections, as did Steiner, by means of circles.

THE AUTHOR.

ZÜRICH, *March 8th,* 1866.

THE AUTHOR'S PREFACE TO THE THIRD
EDITION OF PART I.

THIS new edition exceeds the first in extent by about two-thirds of its number of pages. The most significant changes made in the second edition consisted in the addition of a collection of two hundred and twenty-three problems and theorems. A part of this collection was originally to be found in the appendix to the second volume, but this has been considerably enlarged by the addition of new problems and useful theorems. The first eleven sections of this collection, with the exception of the two upon the principle of reciprocal radii and the ruled surface of the third order, correspond to the lectures with the same headings, and problems and theorems which are comparatively easy to be proved have been selected mainly with a view to furnishing exercise for the student.   I urgently advise every beginner to actually solve the problems of construction graphically, since the comprehension of the Geometry of Position is made very much easier by the free use of pencil and paper.

The last four sections of the problems and theorems contain new investigations which were not found in the first edition, and which in more than one essential feature have been carried out by means of synthetic geometry for the first time so far as I know in this book. In order that the investigations upon self-polar quadrangles and self-polar quadrilaterals and upon linear nets and webs of conic sections might not become too voluminous, I have chosen for them a form of statement as brief as possible. By this means and by the introduction of a single elementary notion, I have been able, within the narrow compass of twenty-one pages, to present the important theories of sheaves, ranges, nets, and webs of conic sections in an entirely new connection. By means of Stephen Smith's theorem, the synthetic proof of which I acquired only after many fruitless attempts, and by the principle of reciprocity, these theories are developed in a manner remarkable for simplicity and clearness.

I have replaced the proof of the fundamental theorem of the Geometry of Position as given by Von Staudt by one free from objections, making use of the remarks in that connection of F. Klein and Darboux (*Math. Annalen*, Vol. xvii.). It was expressly assumed in the second edition by the definition of "correlation," that in two projective primitive forms a continuous succession of elements in the one form corresponds to a continuous succession of elements in the other. This has now been proved upon the basis of Von Staudt's definition of projectivity.

It is due to the kindly co-operation of the new publishers that this book, like its Italian and French translations, is supplied with engravings of the diagrams inserted in the text.

THE AUTHOR.

STRASSBURG IN ALSACE,
September 8th, 1886.

CONTENTS.

INTRODUCTION.

1. Most of my hearers will have heard till now scarcely more than the name of the Geometry of Position; for, unfortunately, the knowledge of this significant creation of recent times, distinguished alike by abundance of contents, clearness of form, and simplicity of development, has been diffused but very little; and notwithstanding the fact that modern geometry must be accounted among the most stimulating branches of mathematical science and admits of many beautiful applications to the technical and natural sciences, it has not as yet found its way generally into the schools. Perhaps, therefore, it will not be amiss if I preface my lectures with a word upon the place which the Geometry of Position occupies among other branches of geometry, and if afterwards I mention some theorems and problems which will serve still further to characterize this science for you.

2. The pure Geometry of Position is mainly distinguished from the geometry of ancient times and from analytical geometry, in that it makes no use of the idea of measurement; in contrast with this feature the ancient geometry may be called the 'geometry of measurement,' or 'metric geometry.' In the pure Geometry of Position nothing whatever is said about the bisection of segments of straight lines, about right angles and perpendiculars, about ratios and proportions, about the computation of areas, and just as little about trigonometric ratios and the algebraic equations of curved lines, since all these subjects of the older geometry assume measurement. In these lectures, however, at the end of each main division, I shall make applications of the Geometry of Position to metric geometry, in which I shall assume a knowledge of planimetry as well as, in a few instances, a knowledge of the sine of an angle. We

A

shall be concerned as little with isosceles and equilateral triangles as with right-angled triangles; the rectangle, the regular polygon, and the circle are also excluded from our investigations, except in the case of these applications to metric geometry. We shall treat of the centre, the axes, and the foci of so-called curves of the second order, or conic sections, only as incidental to the general theory; but, on the other hand, shall become acquainted with many properties of these curves more general and more important than those to which most text-books upon analytical geometry are restricted. We shall be obliged to mark out a new way of approach to the conic sections themselves, since in the Geometry of Position we dispense with the help of the circular cone by means of which the ancient geometers defined these curves, and also of the algebraic equation through which they are viewed by disciples of Descartes. After what has been said it is scarcely necessary for me to mention the fact that no computations will appear in my lectures; only once in a while in the applications to metric geometry shall we employ the sign of equality.

3. Of the knowledge of geometry acquired in the elementary schools, I shall therefore make very little use. On the other hand, a certain skill in producing mental images of geometric forms without pictorial representations would be of great service to you, inasmuch as it will not be practicable for me to illustrate every theorem by diagrams, especially if the theorem refers to a form in space; I shall often be compelled to make demands upon your imagination.

Since the imagination is brought much into play in descriptive geometry, a knowledge of this latter science would likewise stand you in good stead; the converse is equally true, that the Geometry of Position makes an excellent preliminary study for descriptive geometry. And in general, I may say that of all branches of geometry the descriptive is most helpful in facilitating the study of the Geometry of Position—in the first place, because it is very closely related to the latter; and in the second place, because in descriptive geometry relations of magnitude come less under consideration than do the positions of forms relatively to one another or to the plane of projection; this relativity, to be sure, often being defined by the use of circles and right angles.

Above all, however, you will find that perspective or central projection plays an important part in the Geometry of Position, and that many forms of expression used in the latter subject are derived from the former.

4. Pure geometry stands in a certain antithetical relation to analytical geometry on account of its method, which, as you know from the geometry of ancient times, is synthetic. In our study we shall start out with a small number of 'primitive forms'; the simple relations which may be established among these will bring us to the so-called forms of the second order, among which are found the conic sections, and will at the same time permit the principal properties of these forms to be easily recognized. We shall then be able to proceed in the same way from the forms of the second order to still other new forms. During our investigations we clearly must avoid all processes of analysis, that powerful instrument of modern mathematics, since we make no use of measurement, and in order to be able to compute with forms in space, we should first be obliged to express them in numbers, *i.e.* measure them.

On account of its methods pure geometry as distinguished from analytical is often designated by the name 'synthetic geometry.'

5. Since in the pure Geometry of Position metric relations are not considered, its theorems and problems are very general and comprehensive. For example, the most important of the properties of conic sections which are proved in text-books on analytical geometry are merely special cases of theorems with which we shall become acquainted later. A few illustrations will serve more exactly to characterize the material with which we shall be concerned in these lectures.

6. In designing architectural structures, and in drawing generally, it is not infrequent that a solution is required of the problem: "To draw a third straight line through the inaccessible point of inter-section of two (converging) straight lines." Metric geometry fur-nishes us with any desired number of points of such a third line by the aid, for instance, of the property that proportional segments are formed upon parallel lines by any three transversals which meet in one point. The Geometry of Position affords a simpler solution, as follows: Choose some point P outside the two given straight lines a and b (Fig. 1) and pass through it any number of transversals. Then ascertain the point of intersection of the diagonals in each of the quadrangles formed by two of these transversals taken with the lines a and b; all these points of intersection lie upon one straight line which passes through the

intersection of a and b.* The proof follows very simply from the
important harmonic properties of a quadrangle, which may be stated
in the following form:

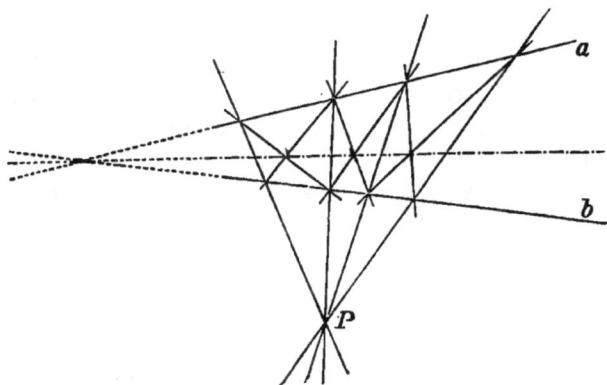

FIG. 1.

If we choose three points A, B, C, upon a straight line (Fig. 2),
and construct any quadrangle such that two opposite sides pass
through A, one diagonal through B, and the other two opposite

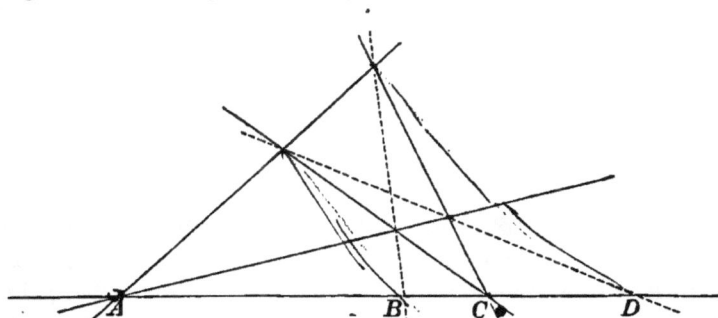

FIG. 2.

sides through C, then the second diagonal meets the straight line
ABC in a perfectly definite fourth point D. The points A, B,

*I strongly recommend the beginner to draw the figure illustrating this
theorem for himself, according to the statement of the text, without first having
seen the one drawn by me, and especially to do so for the subsequent
theorems which are not so simple. A diagram built up by degrees is far
easier to be comprehended, and illustrates most of the theorems to be
represented far better than does one with all its auxiliary lines ready drawn.

C, D are called four harmonic points, and *D* is said to be harmonically separated from *B* by the points *A* and *C.*

By constructing different quadrangles satisfying the stated
conditions, you can easily obtain a confirmation of the fact that
all second diagonals do pass through this fixed fourth point *D.*

The problem stated above can be made use of in surveying when
it is required to extend a straight line beyond an obstacle, for
example, beyond a forest, inasmuch as it affords a means of evading
the obstacle.

7. Of theorems relating to triangles I shall mention only the
following :

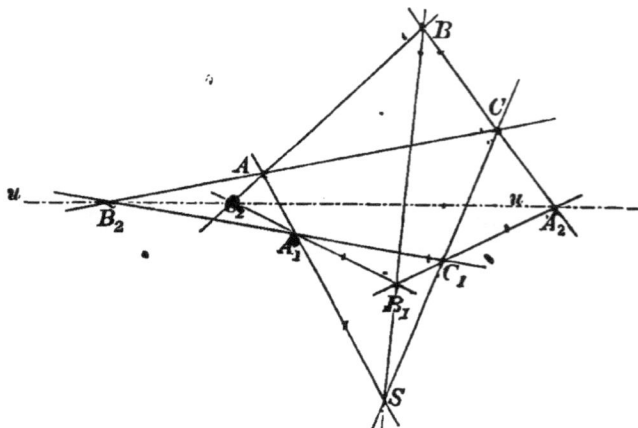

FIG. 3.

If two triangles *ABC* and $A_1B_1C_1$ are so situated (Fig. 3) that
the straight lines joining similarly named vertices, viz. AA_1, BB_1,
CC_1, intersect in one and the same point *S*, then the pairs of
similarly named sides *AB* and A_1B_1, *BC* and B_1C_1, *CA* and C_1A_1,
intersect in three points C_2, A_2, B_2, which lie upon one straight line
u ; and conversely.

The diagram illustrating this theorem is worthy of notice as
representing a class of remarkable configurations characterized by
a certain regularity of form. It consists of ten points and ten
straight lines ; three of the ten points lie upon each of the straight
lines, and three of the ten lines pass through each of the points.

8. Another series of theorems is connected with curves of the
second order or conic sections. You know from analytical geometry,

and will be able later to prove synthetically, that a curve of the second order is completely determined by five points or five tangents. But you also know the difficulty which is met with in the actual computation and construction of a conic section determined in this way. The Geometry of Position establishes two very important theorems concerning curves of the second order, which render it possible for us to construct with ease from five given points or tangents of such a curve any required number of new points or tangents, and so readily to draw the curve itself. Those of you who are already acquainted with these two theorems will remember how much preliminary knowledge is demanded for their proof by analytical methods. The first of these, originally established by Pascal, states that the three pairs of opposite sides of any hexagon inscribed in a curve of the second order intersect in three points which lie upon one straight line; according to the second, which was first enunciated by Brianchon, the three principal diagonals (*i.e.* the straight lines joining pairs of opposite vertices) of any circumscribed hexagon pass through one and the same point. Both theorems may easily be verified in the case of the circle. You will notice that in these theorems nothing is said concerning the size of the conic section, or concerning its centre or its axes or its foci. But just on that account the theorems are of the greatest generality and significance, so that the whole theory of conic sections can be based upon them. In particular, the important problem of drawing a tangent at a given point admits of solution by means of Pascal's theorem, even when the conic section is given by only five of its points, without supposing the whole curve to be drawn.

9. The problem of constructing tangents to a curve of the second order may be solved in many cases by the aid of a theorem which expresses one of the most important properties of these curves, but which is frequently not to be found in text-books on analytical geometry, since its analytical proof is quite complicated and is scarcely capable of setting forth this property in its true light.

Namely, if through a point A (Fig. 4) which lies in the plane of a curve of the second order but not on the curve, secants be drawn to the curve, any two such secants determine four points, as K, L, M, N, upon the curve. Each pair of straight lines, other than the secants, joining four such points two and two, for example, LM and NK, or KM and LN, intersect in a point of a fixed straight line a, which is called the polar of the given point A. If the point

A lies outside the curve, its polar *a* intersects the curve in the points of contact of the two tangents which can be drawn from *A* to the curve; if *A* lies inside the curve, the latter is not cut by the polar. You can apply this theorem in drawing tangents to a conic section from a given point with the use of the ruler only.

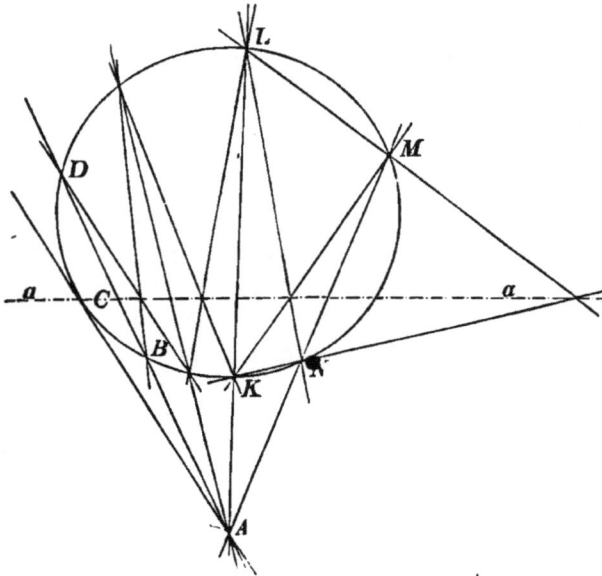

FIG. 4.

Upon any secant passing through *A*, there are four points worthy of notice; first, *A* itself; next, the first point of intersection *B* with the curve; then follows the point of intersection *C* with the polar *a* of *A*; and finally, the second point of intersection *D* with the curve. These four points *A*, *B*, *C*, *D* are harmonic points, and the polar *a* thus contains every point which is harmonically separated from *A* by two points of the curve.

The important theorems relating to centres and conjugate diameters of conic sections are merely special cases of the theorems just mentioned. These latter may easily be extended to surfaces of the second order, since such surfaces are in general intersected by planes in curves of the second order.

From these few examples, which I might multiply indefinitely, you will doubtless have observed how different from the theorems treated in analytical geometry, for instance, are those with which the Geometry of Position is concerned, but certainly they are not less important. I would remind you still further in this connection that analytical geometry seeks to determine the positions of tangents to a conic section, especially by means of the angles which they form with the focal rays, or by means of the intercepts which they determine upon the axes, thus referring the whole matter to metric properties. Of course, reference is made here only of the elements of analytical geometry to which most text-books on the subject are confined, and not of the exceedingly fruitful modern methods, for whose existence we are indebted above all to the ingenious Plücker.

LECTURE I.

THE METHODS OF PROJECTION AND SECTION. THE SIX PRIMITIVE FORMS OF MODERN GEOMETRY.

10. As is well known, the great number of concepts which are advanced in the ancient geometry, in trigonometry, and in analytical geometry are based for the most part upon measurement; accordingly, they can find no place in the pure Geometry of Position. It ought not, therefore, to be surprising that modern geometry has set forth for its purposes a considerable number of characteristic concepts. With these you will be made acquainted in this and in the following lecture, and thereafter they must constantly be employed.

11. The point, the straight line, and the plane are the simple 'elements' of modern geometry.* As a rule, we shall designate points by capital letters, lines by small italics, and planes by Greek letters. Straight lines (or rays, as they will frequently be called) and planes will always be considered as unlimited in extent unless the opposite is expressly stated. We are able to combine these elements into systems by looking upon one of them as the 'base or support' (Träger) of an infinite number of elements of another sort. By this means we arrive at the so-called 'primitive forms' of modern geometry. Before explaining these, I shall, by way of introduction, give a brief account of the important methods of Projection and Section, of which frequent use is made.

12. If we look at an object, say a tree, every (visible) point of it sends to the eye a ray which is called the 'projector,' or the 'projecting ray' of this point. The projector of the whole tree is compounded out of many rays, each of which 'projects' one

---

* It is worthy of notice that no attempt is made to define a 'point,' 'straight line,' or 'plane.' A knowledge of these as fundamental ideas is assumed.—H.

or more points to the eye. If a number of points lie in a straight line not passing through the eye, all their projecting rays lie in that plane which can be passed through the eye and this straight line; every such straight line is projected to the eye by a plane which is called the 'projector,' or the 'projecting plane' of this line. Similarly, a curve is in general projected by a conical surface. We can now intercept, or 'intersect,' the projector of the tree by a plane, each projecting ray being cut in a point and each projecting plane in a straight line. By this means we obtain in the plane, as the 'section' or 'trace' of this projector, a perspective picture, a 'projection' of the tree, and this projection evidently throws the same projector into the eye as the tree itself, and is therefore quite competent to convey a notion of the latter to us. Ordinary photographs of three-dimensional objects are essentially such perspective, plane pictures of the objects.

Upon this kind of projection, which is known by the name of 'central projection,' is based the theory of perspective; and all other varieties of projection which are in use in descriptive geometry may be looked upon as special cases of this one. In orthogonal projection, for example, in order that the projecting rays may be parallel we need only to imagine the eye removed to an infinite distance. The shadows which objects throw upon planes, when they are illuminated from a finite or infinitely distant point, are nothing else than projections of these objects in which the illuminating point takes the place of the eye.

13. A simple example may show how we are able to discover, and at the same time can prove, important theorems through mere visualization, with the help of these methods of projection and section. A system of parallel lines is projected from the eye by planes, all of which intersect in one and the same straight line, namely, in that parallel which passes through the eye; these projecting planes are intersected by an arbitrary picture plane in straight lines, all passing through one point, namely, through the trace of the line in which the projecting planes intersect. Consequently, in the perspective view of a tree or other object the projections of parallel edges converge toward one point, their so-called vanishing point, and only in one particular case, which you will at once recall, are these projections also parallel. We have thus incidentally established and proved a well-known fundamental theorem of central projection.

14. Leaving aside all optical references, let us now further employ the expressions just used, viz. 'projector,' 'ray,' 'to project,' 'to intersect,' etc., where instead of the eye we shall choose an arbitrary point S, and instead of the definite object or tree, an arbitrary system Ω of points and straight lines in space. This system Ω is projected from S by a system of rays and planes, namely, each point by a ray and each ray, not passing through S, by a plane. The point S is regarded as the 'base' of all these rays and planes which together form the projector of the system Ω. If we choose in space an arbitrary system Σ of planes and straight lines, then any new plane ϵ would 'intersect' this in a system of straight lines and points, namely, in general, each plane in a straight line and each straight line in a point. The plane ϵ appears in this case as the 'base' of all these straight lines and points which together make up the 'section' (the 'trace') of the system Σ.

15. We can also project from and make sections by straight lines. Thus every point lying outside a straight line g, together with g, determines a plane, or is 'projected' from g by a plane; and similarly, every plane which does not contain g is 'intersected' by this straight line in a point. The straight line appears, in the first case, as the base of planes which intersect in it; in the second case, as the base of points which lie upon it.

16. Through such considerations as these we obtain the following so-called primitive forms, which occupy an important place in modern geometry.

The totality of points lying upon one straight line is called a 'range of points' (Punktreihe) or a 'line form' (gerades Gebilde); the individual points of the straight line are the elements of the range of points. We consider these points to be rigidly connected with one another, so that their relative positions remain unaltered if the straight line, their base, be moved out of its original position. A portion of a range of points bounded by two points of the range is called a 'segment.'

The totality of rays passing through one point and lying in one and the same plane we shall call a 'sheaf of rays' (Strahlenbüschel). The common point of intersection of the rays is called the 'centre' of the sheaf; the single rays, unlimited on either side of the centre, are the elements of the sheaf. Here again we imagine the elements to be rigidly connected with one another.

Either the centre or the plane in which the rays lie may be

looked upon as the 'base' of the sheaf of rays. A portion· of a
sheaf of rays bounded by two rays of the sheaf as 'sides'· is
called a 'complete plane angle.' This consists of two 'simple'
angles which are vertically opposite to each other. In any sheaf
of rays S (Fig. 5), if four rays a, b, c, d are chosen at random,
then among these there are two pairs
of separated rays, for instance, a and
c are separated from each other by
b and d, so that we cannot pass in
the sheaf from a to c without crossing either b or d.

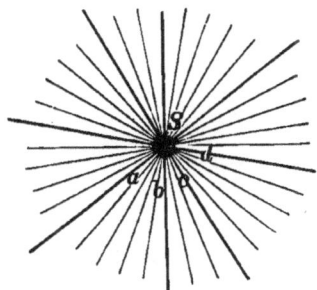

FIG. 5.

The totality of planes, unlimited in
all directions, which pass through one
straight line we shall speak of as a
'sheaf of planes' (Ebenenbüschel) and
the straight line shall be called the
'axis' of the sheaf. As in the range of points, so here, we consider
the elements of the sheaf, that is its planes, to be rigidly connected
in unalterable relative positions. A portion of a sheaf of planes
bounded by two planes as 'faces' is called a 'complete dihedral
angle,' and consists of two 'simple' dihedral angles which are
vertically opposite to each other. Among four planes of a sheaf
two pairs again are separated.

If no confusion is likely to be caused, a form which consists
only of discrete points and the intervening segments of a straight
line will often be called a range of points. In the same way, a form
which comprises only discrete elements and the included angles of
a sheaf will often be spoken of as a sheaf. In doing so we must
constantly bear in mind that, deviating from the ordinary definition,
we have included angles as part of a sheaf.

17. We designate the range of points, the sheaf of lines and
the sheaf of planes as *one-dimensional* primitive forms or primitive
forms of the *first* grade. The elements of a one-dimensional
primitive form, for example the planes of a sheaf, are to be
looked upon as simple elements, *i.e.* they are to be viewed apart
from the forms (geometrical figures, and the like) whose bases they
might be. In the case of the sheaf of rays this view is facilitated
if we distinguish the straight lines whose totality makes up the
sheaf by the name 'rays.' For, by a ray is ordinarily meant a
straight line in itself, viewed apart from the points lying on it

or the planes passing through it. Unfortunately, there is no corresponding second designation for the plane available.

Of the primitive forms of the first grade, moreover, we can imagine any one to be generated by either of the others. Thus a range of points $ABCD$ (Fig. 6) is projected from an outside

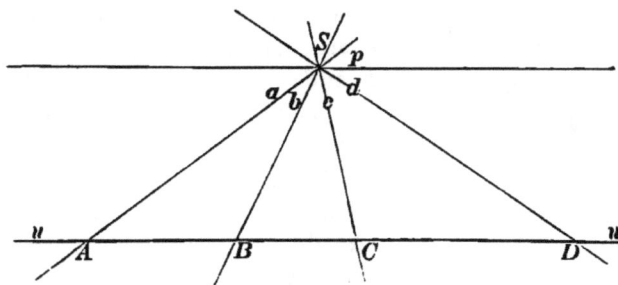

FIG. 6.

point S by a sheaf of rays $abcd$, of which the range $ABCD$ is a section. In the same way the range $ADCB$ is projected by the sheaf $adcb$. A sheaf of planes $\alpha\beta\gamma\delta$ is intersected by any plane not passing through its axis in a sheaf of rays $abcd$ whose centre lies upon the axis; every sheaf of rays is projected from a point not lying in its plane by a sheaf of planes. Finally, every sheaf of planes is intersected by a straight line which does not lie in a plane with its axis, in a range of points; and every range of points is projected from an axis which does not lie in a plane with it, by a sheaf of planes. From these relations it is certainly permissible to characterize the range of points, the sheaf of rays, and the sheaf of planes as primitive forms of the same, namely, of the first grade. For, from what has been said, it is clear that a range of points contains just as many points as a sheaf contains rays or planes.

18. There are two varieties of primitive forms which are said to be of the *second* grade, namely, the plane field and the bundle of rays. The totality of points and lines which are contained in a plane we name a 'plane system or field'; the plane is the 'base' of the system of points and lines. In the plane field there are contained, consequently, not only points and straight lines as elements, but also indefinitely many ranges of points and sheaves of rays; for all the points

lying on a straight line of the field taken together form a range of points, and all the lines of the field passing through one point form a sheaf of rays. The plane field is, therefore, justly characterized as a primitive form of higher grade than the one-dimensional primitive forms.   .

Further, we call the totality of rays and planes which pass through any point in space (as centre) a 'bundle of rays' (Strahlenbündel). In this there are contained as elements not only straight lines and planes, but also indefinitely many sheaves of rays and sheaves of planes. For all planes of the bundle which intersect in one and the same axis form a sheaf of planes; and, in the same way, all straight lines of the bundle which lie in one and the same plane form a sheaf of rays. Thus the bundle of rays is in reality a primitive form of higher grade than the one-dimensional primitive forms.

The term 'bundle,' which is appropriate to denote a multiplicity of higher grade than the term 'sheaf,' was very happily chosen by Von Staudt; we can, however, name the foregoing primitive form 'a bundle of planes' (Ebenenbündel) with the same propriety as a 'bundle of rays,' since it contains planes for elements as well as rays. According also as the points or the straight lines come more into consideration, is the plane field designated as a 'field of points' or a 'field of rays.'

It is scarcely necessary to mention that in the plane field and in the bundle of rays, we imagine the elements of which they consist to be rigidly connected with one another, so that in the bundle, for example, the relative positions of the rays, planes, and sheaves contained therein are unaltered when the centre, which is the base of the bundle, is moved from its original position.

We may further assert that a bundle of rays contains just as many rays and planes as a plane field contains points and rays, and we are therefore wholly justified in considering the two primitive forms as of the same, viz., the second grade, since we can imagine the bundle of rays to be generated from the plane field, and conversely. If we project, for instance, a plane field Σ from an outside point S, so that each point P of Σ is projected by a ray SP of S, and each ray of Σ by a plane of S, then we obtain a bundle of rays S which is called a projector of the field Σ, and of which the field is a section.

To aid your imagination, suppose that Σ is a plane landscape spread out at your feet, unlimited in extent and sparkling in variegated colours, and that the outside point S is your eye. Each point of the landscape, then, sends into your eye a ray of light, each straight line of the landscape a plane of light. If now you consider these rays and planes as unlimited in all directions, you obviously have a bundle of rays as projector of the whole landscape.

We may further conclude that each range of points of the plane field is projected from S by a sheaf of rays, each sheaf of rays by a sheaf of planes, each curve by a conical surface belonging to the bundle ; or, in other words, the projectors of a range of points, a sheaf of rays, and a curve lying in the plane field are respectively a sheaf of rays, a sheaf of planes, and a conical surface in the bundle of rays. Just so, each segment is projected by a plane angle, each plane angle by a dihedral angle, etc.

Conversely, if we consider the bundle of rays as the original form and imagine its centre to be, say, a luminous point which sends out coloured rays on all sides, then the field may be looked upon as a section of this bundle. In this case, each ray of the bundle is cut in a point of the field, each plane in a straight line, each sheaf of rays in a range of points and each sheaf of planes in a sheaf of rays.

19. Finally, there exists a primitive form of the *third* grade, namely, the space system, or unbounded space with all possible points, lines,* and planes in it. The space system contains as elements indefinitely many primitive forms of the first and second grades, since each of its planes is the base of a field, each point the centre of a bundle of rays, each straight line the base of a range of points and at the same time the axis of a sheaf of planes.

20. To each of the six primitive forms which I have just defined there corresponds a distinct geometry. It will be readily conceded that there must be just as complete a geometry for the bundle of rays as for the plane field ; for, corresponding to every plane geometrical figure we immediately construct a form in the bundle by projecting the plane field from an outside point. The theorems which may be enunciated concerning plane figures can be carried

---

* Properly speaking, the space system viewed as consisting of lines is of four dimensions.—H.

over in some manner to their projectors in the bundle of rays. We shall have occasion to make frequent use of this process.

It is more difficult to perceive that there is also a geometry for the one-dimensional primitive forms, for example, for the range of points, *i.e.* the points of a straight line; but I need only to recall the theorem upon harmonic points cited in the introduction, in order to convince you of this fact. The statement I made there was that the position of a fourth harmonic point is determined by three points of a straight line. To show, further, that in reality something of a geometry of one-dimensional primitive forms can be established without the aid of measurement, let me recall the fact that among four rays of a sheaf there are two pairs of rays which are separated by the others.

21. The discussions up to this point make it possible for me now to indicate the principal contents of the Geometry of Position in a very few words. That is to say, the Geometry of Position treats of the six primitive forms mentioned in this lecture and of their mutual relations.

LECTURE II.

22. In the ancient geometry two straight lines are said to be
parallel if they lie in the same plane and have no point in common.
Likewise, two planes or a plane and a straight line are parallel
if no point of the one lies at the same time in the other. Modern
geometry establishes parallelism in a different manner, and it
will be my next object to make you familiar with this modern
conception which Von Staudt has called the perspective view of
parallelism. We are brought directly to this when we derive one
primitive form from another by considering the first a section
or a projector of the second.

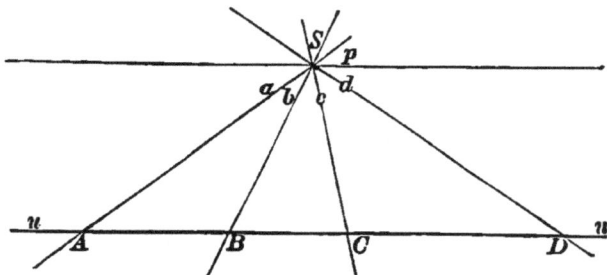

Fig. 6.

If a straight line u (Fig. 6) lies in a plane with a sheaf of
rays S without passing through its centre, then it intersects the
sheaf in a range of points, namely, each ray a, b, c,...of S is
cut in a point A, B, C,...of u. If now by rotating about S in
a fixed sense abc, any ray describes the sheaf S, its trace upon

B

the straight line *u* at the same time describes the range of points *u* in the sense *ABC*. The point of intersection moves from the position *A* farther and farther out beyond *B* until it is lost to view, and then returns, from an infinite distance on the opposite side back to its original position. According to the ancient notion the rotating ray no longer intersects the line *u* in the one particular position *p* in which it is said to be parallel to *u*, and on this account the general statement that every straight line which lies in a plane with *u* has one point in common with it is not allowable. In modern geometry the exceptional case is removed by attributing also to two parallel lines a common point, namely, an infinitely distant point.

23. From the perspective point of view, moreover, any straight line has *only one* infinitely distant point, since, in accordance with one of Euclid's axioms, only one straight line *p* can be drawn parallel to *u* through a given outside point *S*, and to this parallel line is attributed but one point in common with *u*, just as every other ray of the sheaf *S* has only one point in common with *u*.* This conception of parallelism presents distinct advantages over the ancient one in that, first, many theorems can be enunciated in a perfectly general form for which, otherwise, exceptions would need always to be cited, and second, many apparently different theorems can in accordance with this view be comprised in a single statement. You have already become familiar with this idea in analytical geometry. There we are accustomed to call straight lines which lie in one plane parallel if the coordinates of their point of intersection are infinitely great, the point thus lying at an infinite distance.

24. The infinitely distant point of a straight line is approached by a point which moves continually forward upon the line either in the one sense or in the other. Thus the infinitely distant point lies out in both directions † upon the line, or as properly in the one direction as in the other, and the straight line appears to be closed, its extremities meeting in the infinitely distant

---

*The assumption of a single infinitely distant point on a straight line and the definition of parallel lines as lines which intersect in a common infinitely distant point is equivalent to the assumption of Euclid's twelfth axiom. With this axiom as starting point, Euclid proves that one and only one straight line can be drawn through a given point parallel to a given straight line.—H.

† The term 'direction' is used here in its colloquial sense.—H.

point. We are forced to this conclusion as soon as we admit the assumption that every straight line contains one and only one infinitely distant point. We shall see later that the two branches of a hyperbola are to be considered as connected at infinity in just the same way. Analysis leads to similar conclusions, pointing out by frequent examples that we can pass from the positive to the negative not only through zero but also through infinity.

25. Since, then, we can go from one point of a straight line to another by passing over the infinitely distant point of the line, the following statement is true: Among four points of a range there are only two pairs whose elements are separated by the remaining points of the quadruple. This is strictly analogous to the fact that among four elements of a sheaf only two pairs of elements can be so chosen that the elements composing a pair are separated by the other two; and just as a sheaf is divided by two of its elements into two complete angles (these being supplementary angles), so a range of points is divided by two of its points into two segments of which each is called the 'supplement' of the other. One of these two segments contains the infinitely distant point of the line unless this point itself forms one of the boundaries of the segments. In the latter case each of the two segments may be called a *half-ray*.

26. In order to distinguish the infinitely distant point of a straight line from the points of the line which lie in the finite region, we shall call the former an 'ideal' point and the latter 'actual' points. The modern conception of parallelism, explained at the opening of this lecture, might also be characterized as *ideal*. All the parallel lines which may be drawn in a plane in any one direction have but one infinitely distant or ideal point in common, namely, that point which any one of them has in common with each of the others. These parallels may therefore be considered as forming a sheaf whose centre is an infinitely distant point of the plane, and such a system we shall hereafter designate as a sheaf of parallel rays, whenever a distinction from other sheaves of rays is desirable. Likewise under the name 'bundle of parallel rays' are to be comprehended all possible parallel rays in space, having a given direction, together with all planes passing through them. I would remind you at this point that the statements "parallel lines have the same direction" and "parallel lines con-

tain the same infinitely distant point," mean exactly the same thing. Any given direction determines one infinitely distant point, and conversely, each ideal point in space determines a single direction ; moreover, every actual straight line determines both a direction and an infinitely distant point, namely, the ideal point lying upon it.

27. It is assumed of all the infinitely distant points of a plane that they lie in an infinitely distant or ideal line.* This line must be looked upon as a straight line, since it is intersected by every actual straight line of the plane in only one point—the infinitely distant point of that line—while curved lines may have, in common with a straight line, more than one point.

Another reason for this view is the fact that in accordance with the perspective idea two parallel planes must have all their infinitely distant points in common. For, if these planes are cut by any third plane in two actual straight lines, these lines can intersect in no actual point ; they are therefore parallel, since they lie in one plane, and consequently have in common an infinitely distant point of both planes. In this manner it may be shown that every infinitely distant point of the one plane lies also in the other. But since, in general, any two intersecting planes have a single straight line in common, we attribute also to two parallel planes a single common straight line.

28. As it is said of parallel lines that they have the same *direction*, so we say of parallel planes that they have the same *aspect* ; just as, then, in every direction there lies an infinitely distant point, so for every aspect there is an infinitely distant straight line. All parallel planes in space of any one aspect pass through one and the same infinitely distant straight line, namely, through that straight line in which some one of these planes is intersected by each of the others. Parallel planes may therefore be considered as forming a sheaf of planes whose axis is an infinitely distant straight line ; this we shall call a 'sheaf of parallel planes.'

29. Of the infinitely distant points and lines of space it is assumed that they lie in an infinitely distant or ideal surface ;

---

* That is, that the infinitely distant points of a plane form a continuum. If an actual straight line of the plane be rotated about one of its actual points every other actual point will describe a continuous line. The same is assumed of the infinitely distant point.—H.

this surface must be considered plane, since it is intersected by every actual straight line in only one point and by every actual plane in a straight line. The infinitely distant or ideal plane is common to all bundles of parallel rays and to all sheaves of parallel planes, since it passes through the centre of the former and through the axis of the latter.

In the same way, the infinitely distant line in any plane is a ray common to all sheaves of parallel rays lying in that plane, since it passes through the centre of each of them. No definite direction can therefore be assigned to the infinitely distant straight line of a plane, but it possesses the direction (contains the infinitely distant point) of every straight line of the plane.

30. Some light is thrown upon the question of infinitely distant or ideal elements by the relations which may be established between two primitive forms. Two forms are said to be 'correlated' to each other if with every element of the one is associated an element of the other. Two elements of the forms which so appertain to each other are called 'corresponding' or 'homologous' elements. If two primitive forms are correlated to a third, then are they also correlated to each other. For to every element of the third there corresponds an element of each of the other two forms, and these two elements are by this means associated with each other.

31. Two primitive forms of different kinds are correlated to each other in the simplest and clearest manner by making the one a section or a projector of the other. For example, if a sheaf of rays S (Fig. 6) lies in a plane with a range of points u not passing through its centre, we may assign to each ray of the sheaf that point of the range which lies upon it. To the parallel ray p of S corresponds, then, the infinitely distant point of u.

Again, if a plane field Σ is considered as being a section of a bundle of rays S whose centre lies outside Σ, then Σ and S are correlated to each other in such a manner that to each point of Σ corresponds the ray of S passing through it, and to each straight line of Σ the plane of S passing through it. The plane of S parallel to Σ corresponds therefore to the infinitely distant line of Σ and to each ray of S lying in this plane corresponds its infinitely distant point lying in Σ. To each sheaf of planes in S corresponds the sheaf of rays in which it is cut by Σ; the latter is a sheaf of parallel rays if the axis of the sheaf of planes is parallel to Σ. If S is a bundle of parallel rays, its centre lying infinitely distant, then to each actual point of Σ corresponds an actual ray of S, and to each ideal element of Σ an ideal element of S. If Σ is the infinitely distant plane and S a point of the finite region, then to each ray of S corresponds its infinitely distant point; to each plane, its infinitely distant line; to each sheaf of rays, an infinitely distant range of points; and to each sheaf of planes, an infinitely distant sheaf of rays.

32. Two primitive forms of the same kind may be correlated most simply by considering them to be sections or projectors of one and the same third primitive form. Thus, in two sheaves of

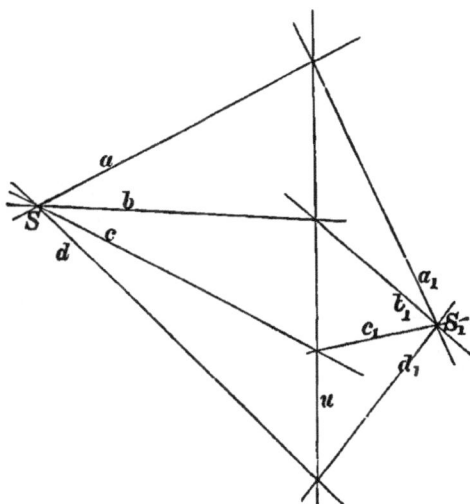

FIG. 7.

rays, or ranges of points, which are sections of one and the same sheaf of planes, those two rays, or points, correspond which lie in the same plane of the sheaf. On the other hand, two sheaves of rays S and S_1 (Fig. 7) can easily be so correlated that they are projectors of one and the same range of points u, *i.e.* so that those pairs of rays a and a_1, b and b_1, c and c_1,... which intersect in points of the range are corresponding rays. If two ranges of points u and u_1 (Fig. 8) lying in one plane be considered sections of one sheaf of rays S, then it is worthy of notice that to the infinitely distant point P (Q_1) of one range corresponds, in general, a point P_1 (Q) lying in the finite portion of the other.

Two plane fields are correlated to each other if they are sections of one and the same bundle of rays. For example, an extended plane landscape and the perspective picture of it which we obtain by intercepting its projector from the eye by any plane, a vertical one, say, are so correlated that those points of the landscape and picture correspond which lie upon the same ray of the bundle having the eye for centre, that is, any two points of the landscape and picture correspond which are found in a straight line with the eye. To each straight line of the landscape corresponds a straight line ȯf the picture, and the two straight lines lie in a plane with the eye. To the infinitely distant straight line of the landscape

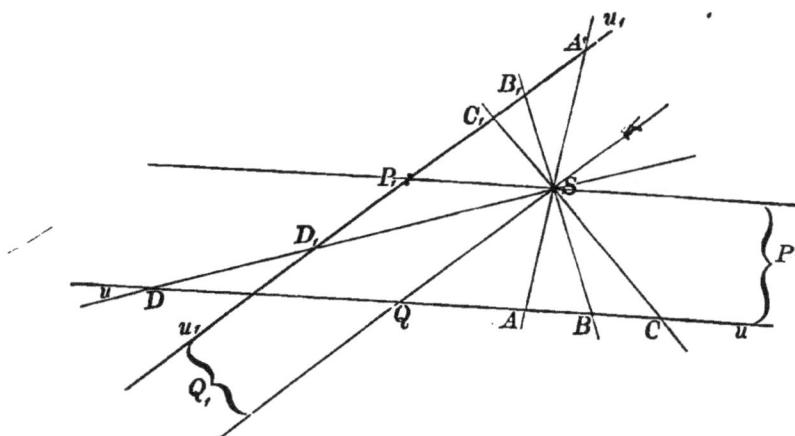

FIG. 8.

(the horizon) corresponds, in general, an actual straight line of the picture, another reason or warrant for considering the infinitely distant line of a plane to be a straight line. Of two plane fields correlated to each other in this manner we appropriately say that one is a 'projection' of the other, and the centre of the bundle which is at once a projector of both fields is called the 'centre of projection.' If the centre lies infinitely distant, the bundle is a bundle of parallel rays, in which case the process of projection becomes the ordinary parallel projection of descriptive geometry.

Two bundles of rays are correlated to each other if we conceive them to be projectors of one and the same plane field. Each ray of the one bundle intersects, then, the corresponding ray of the other in a point of the field; likewise, every two homo-

logous planes of the bundles have a straight line of the field as
their intersection.   The projectors of a plane landscape viewed
from two different points constitute such bundles.

33. I must leave the immediate discussion of the correlation of
primitive forms to each other for the time being to your own
efforts ; I remark, however, that the primitive forms may be correlated in other and more complicated ways.   For example, it is
possible to correlate two plane fields to each other by correlating
them to one and the same third field.   Referring again to an
illustration which has been frequently used, you may imagine
perspective pictures of a landscape to be constructed from two
different centres of projection.   Two such pictures or plane fields,
then, are correlated to each other, since each is correlated to the
landscape ; and clearly, two points of these correspond if they are
projections of the same point of the landscape.   A straight line
of the one picture would correspond always to a straight line of
the other.   But such plane fields, in general, have no longer the
particular position with respect to each other which was previously
discussed, that namely, in which corresponding lines lie in a plane
and the joining lines of pairs of homologous points all intersect
in one point.   Later, we shall have to investigate more minutely
the mutual relations of the two fields so correlated to each other.
Two fields may be so correlated that each straight line of the
one corresponds to a curve in the other, or that to each point
of the one field corresponds a straight line of the other, and
conversely, to each straight line of the former, a point of the
latter.   For the present, however, it is left to your own ingenuity
to work out these more diverse relations of primitive forms.

LECTURE III.

THE PRINCIPLE OF RECIPROCITY OR DUALITY. SIMPLE AND COMPLETE *n*-POINTS, *n*-SIDES, *n*-EDGES, ETC.

34. Before developing further the correlations which may be established among the primitive forms of modern geometry, I must make mention of a geometrical principle which will occupy an important place in these lectures. This principle very greatly simplifies the study of the Geometry of Position, in that it divides the voluminous material of the subject into two parts, and sets these over against each other in such a way that the one part arises immediately out of the other. This principle of reciprocity or duality as it is called was first established in an elementary way by Gergonne,* Poncelet † having previously shown by means of the polar theory that to every figure in space there can be constructed one which corresponds to it in the dualistic sense.

35. Although the principle of duality cannot be generally applied in metric geometry, yet there are many theorems in metric geometry which point directly toward this principle, and which I need only to call to mind in order to make you aware of its existence. In three-dimensional space, the point and the plane stand opposed to each other as 'reciprocal elements,' so that every theorem of the Geometry of Position finds its complement in another which we may deduce from the first by interchanging the terms 'point' and 'plane,' and hence also 'range of points' and 'sheaf of planes,' 'segment' and 'dihedral angle,' etc. Ordinarily we shall write two such 'reciprocal' theorems side by side as the two members of one theorem. For example:

---

* Gergonne, Annales de Mathématiques, T. XVI., 1826.
† Poncelet, Traité des propriétés projectives des figures, Paris, 1822.

Two points A and B determine a straight line AB, namely, the line joining them.

A straight line a and a point B not lying upon it determine a plane aB which passes through both.

Three points A, B, C, which do not lie in one straight line, determine a plane ABC (the joining plane).

Two straight lines a and b, which have one point in common, lie in one plane ab.

Two planes a and β determine a straight line $a\beta$, namely, their line of intersection.

A straight line a and a plane β not passing through it determine a point $a\beta$ which lies upon both.

Three planes a, β, γ, which do not pass through one straight line, determine a point $a\beta\gamma$ (the point of intersection).

Two straight lines a and b, which lie in one plane, have one point ab in common.

36. Incidentally you will notice from these few theorems how useful the introduction of the infinitely distant or ideal elements into geometry proves to be. Without these we should not have been able to enunciate all the above theorems in a general form, but must have called particular attention to special cases as exceptions. The first theorem on the right for example would have read : " Two planes a and β either determine a straight line $a\beta$, or they are parallel," while from the new point of view a straight line is also determined in the latter case, namely, the infinitely distant straight line of the planes. In the same way we should have been obliged to distinguish several cases of the first theorem on the left, according as the two given points A and B are actual points or not. It would have read : " A straight line is determined by two given (actual) points or by one point and a given direction " ; from the perspective point of view, however, the latter case is included in the former, since among the given points infinitely distant points may also be considered. You can yourselves easily make similar observations upon each of the other theorems.

37. For the sake of brevity we shall call two straight lines 'incident' if they intersect ; a straight line, or plane, and a point are 'incident' if the latter lies in the former ; and finally, a ray, or point, and a plane, if the latter passes through the former. Straight lines not incident are called 'gauche.'

38. The foregoing theorems lead now to the following problems, the solutions of which we shall always in future consider possible :

Through two points to pass a straight line.

To find the line of intersection of two planes.

Through a straight line and a point not incident with it to pass a plane.	To find the point of intersection of a straight line and a plane not incident with it.
Through three points to pass a plane.	To find the common point of three planes.
Through two incident straight lines to pass a plane.	To find the common point of two incident straight lines.

39. For the sake of practice, I shall cite a few double theorems which are in frequent use. I strongly urge you to deduce for yourselves, from the one half of each of these, the other reciprocal half.

If four points A, B, C, D are given, and the lines AB and CD intersect, then the four points lie in one and the same plane, so that the lines AC and BD, as well as AD and BC, must also intersect.	If four planes α, β, γ, δ are given, and the lines of intersection $\alpha\beta$ and $\gamma\delta$ intersect, then the four planes pass through one and the same point, so that the lines $\alpha\gamma$ and $\beta\delta$, as well as $\alpha\delta$ and $\beta\gamma$, must also intersect.

If of any number of straight lines each intersects every other one while they do not all

pass through one point, then they must all lie in one plane.	lie in one plane, then they must all pass through one point.

Frequently, a theorem is reciprocal to itself as when point and plane appear in it symmetrically; for example, the problem: In a plane, to draw through a given point a straight line to meet a given straight line which neither lies in the given plane nor passes through the given point.

Here two solutions stand reciprocally opposed to each other:

We may either join the point of intersection of the straight line and plane, with the given point;	or may pass a plane through the straight line and the given point, and find its line of intersection with the given plane.

The following problem may easily be reduced to the one just stated :

Through a given point to draw a straight line which intersects two given straight lines not lying in one and the same plane with the given point.	In a given plane to draw a straight line which intersects two given straight lines not having one and the same point in common with the given plane.

Pass a plane through the given point and one of the given straight lines,	Determine the point of intersection of the given plane with one of the given straight lines,

and the problem becomes identical with the preceding one.

The problem, "To draw a straight line which intersects three given straight lines," is likewise reciprocal to itself. We may either choose a point in one of the straight lines or pass a plane through one of them, and then under the conditions of the preceding problem find a straight line which passes through this point, or lies in this plane, and intersects the other two given straight lines.

40. The primitive forms can also be opposed reciprocally to one another; for example, the plane field and the bundle of rays are clearly reciprocals, since their bases, namely, the plane and the point, are reciprocal elements.

Hence,	are reciprocal to
the point, the range of points, the ray considered as joining two points, the sheaf of rays, etc., in the plane field,	the plane, the sheaf of planes, the ray considered as the intersection of two planes, the sheaf of rays, etc., in the bundle of rays.

The observation will press itself upon you here, as in many previous theorems, that in space of three dimensions the straight line (or ray) is reciprocal to itself. In reality, the straight line occupies an intermediate position between the reciprocal elements point and plane.

41. The following serves as an example of a double theorem in which the plane field and the bundle of rays are opposed to each other as reciprocal forms:

If two plane fields are correlated to each other by considering them as sections of one and the same bundle, then pairs of corresponding elements (points or lines) of the fields lie upon one and the same element (ray or plane) of the bundle.	If two bundles are correlated to each other by considering them as projectors of one and the same field, then pairs of corresponding elements (rays or planes) of the bundles pass through one and the same element (point or straight line) of the field.
The line of intersection of the two planes coincides with its corresponding line, and hence corresponds to itself. The same is true of each point found in this line. The two plane fields there-	The common ray of the bundles, *i.e.* the ray which joins their centres, coincides with its corresponding ray, and hence corresponds to itself. The same is true of each plane

fore have a 'self-corresponding' range of points.

passing through this ray. The two bundles therefore have a 'self-corresponding' sheaf of planes.

42. If two forms are correlated to each other, and an element of one coincides with (*i.e.*, is identical with) its corresponding element in the other, then we say that this element (double element) is a *self-corresponding element* in the two forms.

43. As the point and the plane are reciprocal elements in space of three dimensions, so in space of two dimensions, the point and the straight line, also the range of points and the sheaf of rays, the segment and the angle, are opposed to each other as reciprocal forms ; and similarly in the bundle of rays, the ray and the plane, the sheaf of rays and the sheaf of planes, etc., are reciprocal forms. For example :

(a_1) Any two points of a plane determine a straight line.

(a_2) Any two straight lines of a plane determine a point.

(a_3) Any two rays of a bundle determine a plane.

(a_4) Any two planes of a bundle determine a ray.

A plane curve may be looked upon

(β_1) As the aggregate of the points lying upon it.

(β_2) As the aggregate of the straight lines (tangents) enveloping it (Fig. 9).

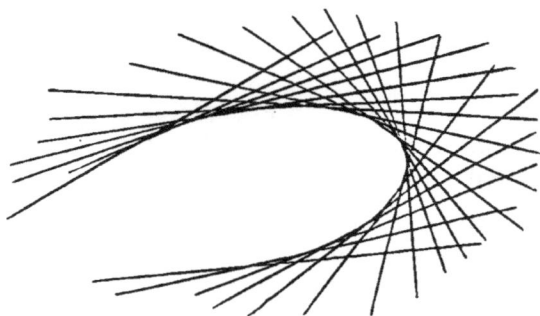

·FIG. 9.

And you will find that in the modern geometry the latter conception is brought into use just as frequently as the former. In the same way, a conical surface (in the bundle of rays) can be looked upon

(β_3) As the aggregate of the rays lying upon it.

(β_4) As the aggregate of the planes (tangent planes) enveloping it.

44. Of four theorems related to one another as are those of the last article, the two relating to the bundle of rays can always be deduced from the other two by projecting the plane field from any centre. As a rule, therefore, I shall in future state only the two planimetric theorems, and will leave you to seek out the others for yourselves. In space, where point and plane are reciprocal to each other, the first and last (a_1 and a_4), also the second and third (a_2 and a_3) of any four such theorems offset each other as reciprocal theorems.

45. The principle of reciprocity will become clearer and more familiar to you in the course of our investigations; but only after a series of developments upon the one-dimensional primitive forms, can I demonstrate that it has general validity in the Geometry of Position, or that in reality to every theorem there corresponds a reciprocal theorem. In the meantime, I shall so adjust my lectures that theorems associated reciprocally with each other will be placed side by side, and I shall so carry out their demonstrations that the dualism will stand out very clearly. To this end, it is necessary that I should develop beforehand some reciprocal ideas, and in particular modify some of those geometrical notions which you have brought over with you from metric geometry.

46. I refer here particularly to the conception of the polygon. In modern geometry we understand by a 'simple plane *n*-point' not as a rule a portion of the plane which is bounded on all sides by *n* intersecting straight lines, but a set of *n* points of a plane and the *n* straight lines or sides, each of which joins two consecutive points or vertices. We look upon the points as being arranged in a definite order, and specify that no three consecutive points shall lie in one straight line.

The simple *n*-point might also be named a 'simple *n*-side,' since a simple *n*-side is a set of *n* straight lines of a plane (the sides of the figure), and the *n* points in which two consecutive sides intersect.

The *n*-point and *n*-side are reciprocal figures. To the lines joining two non-consecutive vertices (*i.e.* to the diagonals) of a simple *n*-point, the points of intersection of non-consecutive sides in the *n*-side are reciprocal, each to each.

47. In metric geometry where by an *n*-point is meant a portion of a plane enclosed by *n*-sides, the re-entrant *n*-point, such as the pentagon *ABCDE* (Fig. 10), or the hexagon *ABCDEF* (Fig. 11) is generally excluded from consideration.

The n-point and the n-side of modern geometry give very little occasion for distinction between re-entrant figures and others, since the sides are supposed unlimited in extent. We may in fact call any two of the $2n$ elements (vertices and sides taken together) of a simple n-point or n-side, 'opposite' elements, which are separated from each other by half the number of remaining elements; consequently, the mth and $(n+m)^{th}$ elements, these being reckoned from any one element round the figure in either order, are opposite to each other.

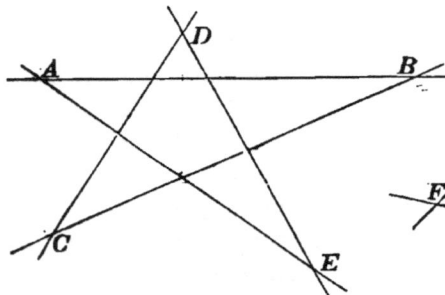 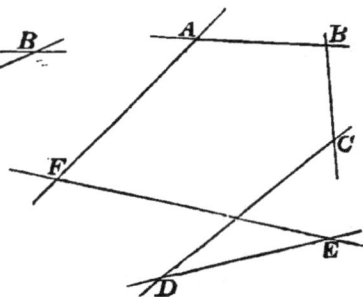

FIG. 10.   FIG. 11.

For example, in the pentagon *ABCDE* (Fig. 10) a vertex and a side lie opposite to each other in pairs, namely, the vertex *A* and the side *CD*, *B* and *DE*, *C* and *EA*, etc.; in the hexagon or hexalateral *ABCDEF* (Fig. 11), on the other hand, the vertices in pairs, and the sides in pairs, as for instance the vertices *A* and *D*, the sides *AB* and *DE*, the vertices *B* and *E*, etc., are opposite elements.

48. Modern geometry, however, deals not only with 'simple n-points and n-sides,' but also with 'complete n-points and n-sides,' and in these figures the principle of reciprocity may again be distinctly recognized. We define as follows:

A complete plane n-*point:* a set of n points of a plane together with all straight lines (sides) joining them two and two, or what is the same thing, a *simple* n-point together with all its diagonals.	*A complete plane* n-*side:* a set of n straight lines of a plane together with all their points of intersection (vertices), or what is the same thing, a *simple* n-side together with all the points of intersection of its sides.

In these definitions it is assumed that no three vertices of the
n-point lie upon the same straight line, and that no three sides
of the n-side pass through the same point.

In each vertex of the complete n-point, $(n-1)$ sides intersect. These pass through the remaining $(n-1)$ vertices (each of them through a second vertex). Hence the total number of sides of the complete n-point is $\frac{1}{2} n (n-1)$.

Upon each side of the complete n-side lie $(n-1)$ vertices; through these pass the remaining $(n-1)$ sides (through each vertex a second side). Hence the total number of vertices of the complete n-side is $\frac{1}{2} n (n-1)$.

49. It is readily seen that many simple n-points and n-sides
are contained in the complete figures whenever n is greater than
three. For example:

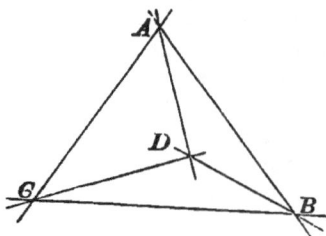

FIG. 12.   FIG. 13.

A complete quadrangle $ABCD$ (Fig. 12) has six sides; any two of these sides as AB and CD, or AC and BD, or finally, AD and BC, which do not pass through one and the same vertex are 'opposite sides' of the quadrangle, so that in a quadrangle there are three pairs of opposite sides. Moreover, the complete quadrangle contains three simple quadrangles $ABCD$, $ACDB$, and $ADBC$, the sides of each consisting of two pairs of opposite sides of the complete figure.

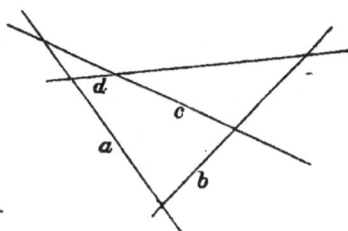

A complete quadrilateral $abcd$ (Fig. 13) has six vertices; any two of these as ab and cd, or ac and bd, or finally, ad and bc, which do not lie upon one and the same side are 'opposite vertices' of the quadrilateral, so that in a quadrilateral there are three pairs of opposite vertices. Moreover, the complete quadrilateral contains three simple quadrilaterals $abcd$, $acdb$, and $adbc$, the vertices of each consisting of two pairs of opposite vertices of the complete figure.

50. The forms in the bundle of rays which correspond to these
plane figures are most easily obtained by projecting the latter from
a point lying outside their plane. Each plane n-point gives rise by
projection to an n-edged figure, or, more briefly, to an ' n-edge,'
and each plane n-side, to an n-faced figure, or an ' n-face.'

Accordingly,

A 'complete n-edge' is a set of n rays of a bundle, together with all planes (faces) passing through them, two and two, assuming that no three of the n rays or 'edges' lie in one plane.

A 'complete n-face' is a set of n planes of a bundle, together with all their lines of intersection (edges), assuming that no three of the n planes or 'faces' pass through one and the same ray.

It would be an easy matter for you to define the 'simple n-edge' and 'simple n-face,' and to develop properties of these forms in the bundles analogous to those of the corresponding plane figures.

51. I shall conclude this series of definitions with those of the analogous space configurations.

A 'complete three-dimensional n-point' consists of n points (vertices) of which no four lie in one plane, the straight lines each of which joins two of the n points, and the planes each of which passes through three of the n points.

A 'complete three-dimensional n-face' consists of n planes (faces) of which no four pass through one point, the straight lines (edges) in each of which two of the n planes intersect, and the points (vertices), in each of which three of the n planes intersect.

I leave the determination of the number of edges and faces of a three-dimensional n-point, as also the edges and vertices of a three-dimensional n-face, to your own inquiry. I remark, however, that the three-dimensional tetragon and the tetrahedron do not differ from each other any more than do the triangle and the trilateral in the plane. That the principle of reciprocity is applicable to the tetrahedron the following theorems among others will show :

The four vertices and six edges of a tetrahedron are projected from any point which lies in none of its faces by the four edges and six faces of a complete four-edge.

The four faces and six edges of a tetrahedron are intersected by any plane which passes through none of its vertices, in the four sides and six vertices of a complete quadrilateral.

You will here observe that in space of three dimensions the complete plane n-point is reciprocal to the complete n-face in a bundle, and the complete plane n-side to the complete n-edge, since point and plane are reciprocal elements.

c

THE CORRELATION OF COMPLETE *n*-POINTS, *n*-SIDES, AND
n-EDGES TO ONE ANOTHER.   HARMONIC FORMS.

52. In my lectures thus far I have sought to solve but one of
the problems lying before me, namely, to make you acquainted
with the most important concepts peculiar to modern geometry.
I have no doubt that you have many times wearied of this multitude
of definitions following in quick succession, but it was necessary
to place these before you in a connected form, so that later we
might bring to light with less interruption the rich treasures which
the Geometry of Position affords.

Let us now proceed to the first real theorem of modern geometry.
The very simple propositions heretofore stated have been mentioned
as occasion might offer, more with a view to familiarizing 🔲 with
the new concepts and for completeness than because they were
all necessary for the establishment of our science.

I shall first call particular attention to the theorems upon harmonic
points, rays, and planes ; in a word, to the theorems upon harmonic
forms in general, which I shall now develop as essentially fundamental in the Geometry of Position.

53. The properties of harmonic forms, of which mention was
made in the Introduction, can be proved most simply by making
use of some elementary theorems upon the correlation of 🔲 points,
n-sides, and *n*-edges to one another.

In a way similar to that by which we have already correlated the
primitive forms, we can associate in certain figures, to each vertex,
side, or edge of one a corresponding element of another. One
quadrangle, for example, can be correlated to a second by associat-
ing with each vertex of the first a vertex of the second ; and in

consequence of this, to each side of the first there will correspond a side of the second.

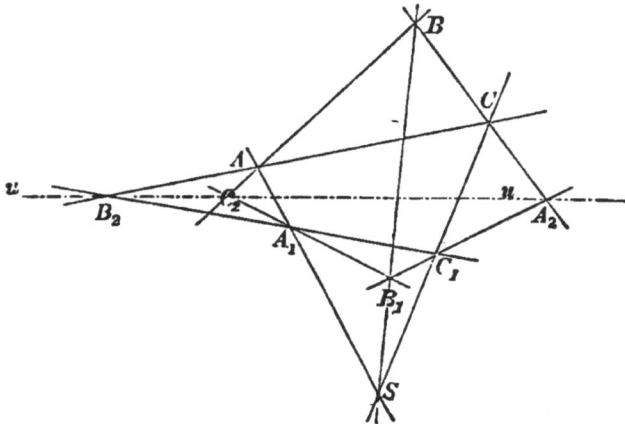

FIG. 13.

We may now state the following self-evident theorem :

If two correlated triangles ABC and $A_1B_1C_1$ (Fig. 3) lie in different planes, and each of the three pairs of homologous sides AB and A_1B_1, AC and A_1C_1, BC and B_1C_1, intersect (necessarily upon the common line u of the planes of the two triangles), then the planes of the three pairs of corresponding sides determine a three-edged figure, of which the two triangles are sections. The joining lines AA_1, BB_1, and CC_1 of the pairs of homologous vertices intersect therefore in one point, namely, in the vertex S of the three-edge.

If two correlated three-edged (or three-faced) figures belong to different bundles, and each of the three pairs of homologous edges intersect, then the three points of intersection determine a triangle of which the two three-edges are projectors. The lines of intersection of the three pairs of homologous planes (faces) of the three-edges lie therefore in the plane of this triangle, whose sides they form.

54. It would be an easy matter for you to enunciate the converse of either half of this double theorem. By the help of these we find that

If two complete quadrangles $ABCD$ and $A_1B_1C_1D_1$ (Fig. 14) lying in different planes whose line of intersection u passes through

If two complete four-faced figures belonging to different bundles of rays whose common ray lies in none of the eight faces, are cornone of the eight vertices are cor-related to each other, and five sides a, b, c, d, e of the one quadrangle intersect (upon u) the correspond-ing sides a_1, b_1, c_1, d_1, e_1 respectively of the other, then are the two quadrangles sections of one and the same complete four-edge, and therefore their remaining two sides f and f_1 also intersect upon u.

related to each other, and five edges of the one intersect the corresponding edges of the other, then are the two four-faced figures projectors of one and the same complete plane quadrilateral, and therefore their two remaining edges also intersect.

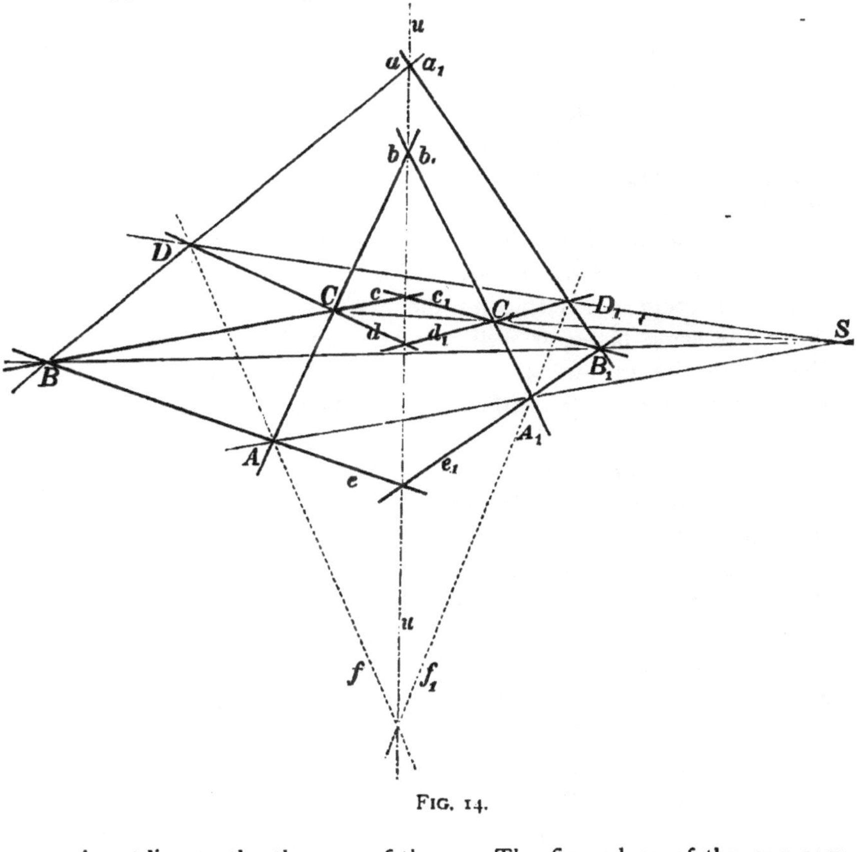

FIG. 14.

According to the theorem of the last article the lines AA_1, BB_1, CC_1, and likewise the lines DD_1, BB_1, CC_1, intersect in one point ; the straight lines AA_1 and DD_1 there-fore meet in the point of intersection

The five edges of the one com-plete four-face which are intersected by the corresponding edges of the other determine in that four-face two three-faced figures, the faces of each of which are intersected by

S of BB_1 and CC_1, the vertex of the four-edge mentioned in the theorem; and since the straight lines f and f_1 lie in the plane determined by AA_1 and DD_1, they must intersect.

the homologous faces of the other in the three sides of a triangle But these two triangles have two sides in common. They lie therefore in one plane, and determine the plane quadrilateral, of which the given four-faces are projectors.

55. In order not to become too profuse I shall at this point drop the investigations of the right-hand column and only make use of a result obtained in the left-hand column in establishing the theory of harmonic elements. But even so we shall very soon reach new theorems which offset each other in just the same way as do those already denoted as reciprocals.

56. We have just now found that—

If, in two complete quadrangles which are correlated to each other, five pairs of homologous sides intersect in points of a straight line u *which passes through none of the eight vertices, then the sixth pair also intersect in a point of this straight line.*

This theorem holds true for the case in which the quadrangles lie in the same plane, as well as when they lie in different planes. For if they lie in the same plane, we can immediately reduce to the case already treated either by rotating one of the quadrangles about the line u, out of the given plane, or by projecting it from an arbitrary centre upon a second plane through u. In either case it happens that through the point of intersection of u with the sixth side of the one quadrangle, the sixth side of the other quadrangle also passes. It may be remarked incidentally that if u is an infinitely distant straight line our theorem would read:

If, in two complete quadrangles which are correlated to each other, five pairs of homologous sides are parallel, then the remaining two sides are also parallel.

57. We may now announce the following definition:

Four points A, B, C, D *of a straight line are called 'harmonic points' (and form a harmonic range of points) if they are so situated that in the first and in the third of them a pair of opposite sides of a quadrangle may intersect, while through the second and fourth points the two diagonals of the quadrangle pass.*

From what has already been said the following important theorem immediately presents itself:

Three points A, B, C *of a straight line and the order of their succession completely determine the fourth harmonic point* D.

For example, we find D by constructing any quadrangle $KLMN$ (Fig. 15), of which a diagonal LN passes through the second point B, two opposite sides KL and MN intersect in the first point A, and the other two opposite sides LM and NK intersect in the third point C; the second diagonal KM will then determine D.

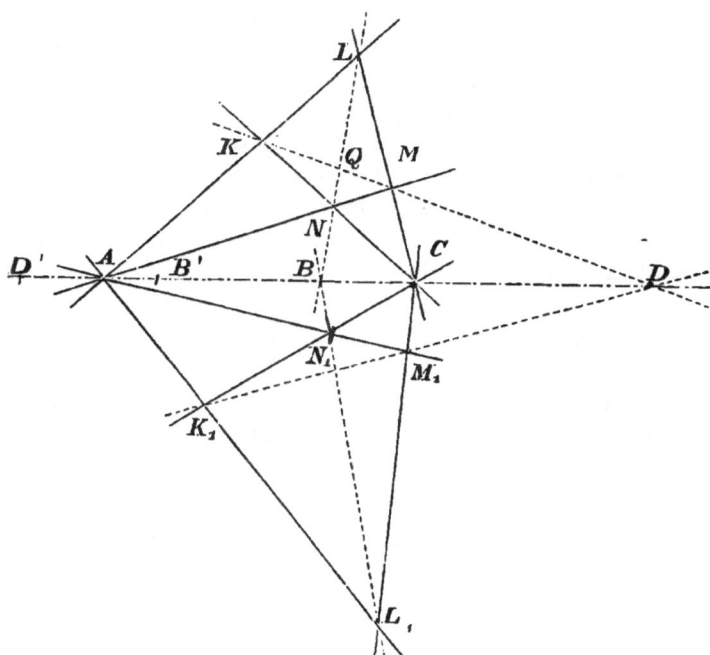

FIG. 15.

If we construct another quadrangle $K_1L_1M_1N_1$ which is related to the points A, B, C in a manner similar to that of the quadrangle $KLMN$, then, in accordance with the theorem of the preceding article, its second diagonal K_1M_1 (being sixth side of the complete quadrangle $K_1L_1M_1N_1$) must also pass through D, the point of intersection of KM and ABC.

The points B and D through which the diagonals pass, are separated from each other by the points A and C, in which pairs of opposite sides intersect, and are therefore said to be 'harmonically separated by A and C.'

58. That the points B and D are in reality *separated* by the points A and C may be demonstrated as follows:

If we project the points A, B, C, D from an arbitrary centre upon a different straight line, then the projectors, and consequently also the projections, of one pair of separated points are separated from each other by the projectors (and also the projections) of the other pair. If now Q (Fig. 15) is the point of intersection of the diagonals KM and LN of the quadrangle $KLMN$, then is $KQMD$ a projection of $ABCD$ from the point L, and $MQKD$, a projection of the same range from the point N. If A were not separated from C, but from B, say, by the remaining two points, then, in the one projection, Q would be separated from K, but in the other projection, from M, which is impossible, since Q can be separated from only one of the three points K, M, D, by the other two. If again A were separated from D, then would D be separated from K and at the same time from M, which likewise is impossible. Consequently A must be separated from C by the points B and D.

59. From a point not lying in the plane of the quadrangle (from your eye for instance), a complete quadrangle is projected by a complete four-edge, the harmonic range of points by a sheaf of four rays which shall be called 'four harmonic rays,' or a 'harmonic sheaf of rays.' These have the property of being intersected by any plane not passing through the centre, in four harmonic points A_1, B_1, C_1, D_1. For every such plane cuts the complete four-edge in a quadrangle of which two opposite sides intersect in A_1, two others in C_1, and whose remaining two sides pass respectively through B_1 and D_1.

60. If four harmonic points are projected from an axis which does not lie in a plane with them, we obtain 'four harmonic planes,' or a 'harmonic sheaf of planes.' Any fifth plane which contains the four harmonic points intersects the four harmonic planes in harmonic rays; and it follows that this is equally true of any intersecting plane whatsoever, which does not pass through the axis of the harmonic sheaf of planes. For any such plane cuts the four harmonic rays of the first intersecting plane in four harmonic points through which its own lines of intersection with the harmonic planes pass. From this it follows further that any straight line gauche to the axis meets four harmonic planes in harmonic points.

In the same way a harmonic sheaf of lines is projected from a point not lying in its plane by a harmonic sheaf of planes. In general,

Four harmonic points are projected from any straight line by four harmonic planes, and from any point by four harmonic rays.	Four harmonic planes are cut by any straight line in four harmonic points, and by any plane in four harmonic rays.

Four harmonic rays are—

Projected from any point by four harmonic planes.	Cut by any plane in four harmonic points.

These several statements may be combined into a single statement of great importance :

From any harmonic primitive form, after projection and section there results always another harmonic primitive form.

At the same time you perceive that by three elements of a one-dimensional primitive form the fourth harmonic is completely determined if it is known from which of the three the fourth is separated. For, if the given elements are three points of a straight line, then the complete quadrangle yields the fourth harmonic point. If, on the other hand, the given elements are rays or planes of a sheaf, these may be sectioned by a straight line, and the fourth harmonic point to the three intersection points may be found. Through this point passes the desired fourth element of the harmonic sheaf. By this means the problem, " From three elements of a one-dimensional primitive form to construct the fourth harmonic," is at once solved.

61. The correctness of the following theorems will be evident to you at a glance :

If three planes α, β, γ, of a sheaf be cut by arbitrary transversals, and upon each transversal is found the fourth harmonic to the three intersection points, this being separated from the point of intersection with β in each case, then all these fourth harmonic points lie in a plane δ, which is the fourth harmonic plane to α, β, γ, separated from β.	If three points A, B, C, of a range be projected from arbitrary axes, and for each axis is found the fourth harmonic to the three projecting planes, this being separated from the plane passed through B in each case, then all these fourth harmonic planes pass through one point D, which is the fourth harmonic point to A, B, C, separated from B.

To these two theorems, which offset each other reciprocally in
space of three dimensions, you will easily be able to state two
corresponding ones for the plane. In them a sheaf of rays takes
the place of the sheaf of planes. Analogous theorems may also
be stated for the bundle of rays.

62. In the definition of harmonic points A, B, C, D, by means
of the quadrangle $KLMN$ (Fig. 15), we have made a distinction
between the two points A and C, in which the opposite sides of
the quadrangle intersect, and the two remaining points B and D,
through which the diagonals pass. It may be shown, however,
that the two pairs of separated points in the harmonic range play
exactly the same part. In the first place, it is evident that of four
harmonic points two separated points may be interchanged without
destroying the harmonic relation, *i.e.* if $ABCD$ is a harmonic
range of points, the same is true of $ADCB$, $CBAD$, $CDAB$;
for in each of these ranges two opposite sides of the quadrangle
$KLMN$ pass through the first, and two through the third, point,
while the diagonals pass through the second and fourth points. If

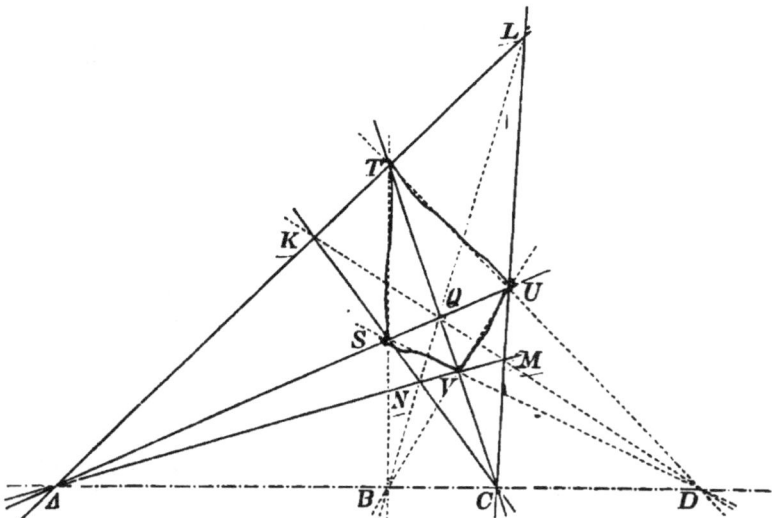

Fɪɢ. 16.

now through the point of intersection Q of the diagonals (Fig. 16)
the straight lines AQ and CQ be drawn, they determine upon the
sides NK, KL, LM, and MN, respectively, four new points
S, T, U, and V. Of the lines ST, TU, UV, and VS, which

are seen to be second diagonals of the quadrangles $KSQT$, $LTQU$, $MUQV$, and $NVQS$, two opposite ones pass through B and the remaining two through D. We thus obtain a quadrangle $STUV$, of which the pairs of opposite sides pass through B and D, and the diagonals through A and C. Hence, in a harmonic range of points, the two pairs of separated points can also be interchanged without destroying the harmonic relation. That is,

If ABCD *is a harmonic range of points, not only are* ADCB, CBAD, *and* CDAB *likewise harmonic, but so also are* DCBA, DABC, BCDA, *and* BADC.

This theorem of course holds good also for harmonic rays and planes, which we have already defined by reference to harmonic points.

63. We say of two separated elements of a harmonic form that they are 'harmonically separated' by the remaining two elements, or are 'conjugate' to each other. For brevity and simplicity of expression we shall frequently say that two elements of a form are harmonically separated by two other elements not belonging to that form, if, by means of the latter elements, two belonging to the form are determined by which the first two elements are harmonically separated. Two points A and C, for example, are said to be harmonically separated by two planes β and δ, if these cut the straight line AC in two points B and D such that $ABCD$ is a harmonic range ; and similarly β and δ are said to be harmonically separated by A and C if they are harmonically separated by the two planes which project the points A and C from their line of intersection.

As an illustration of this mode of expression the following double proposition for forms in a plane may be cited :

From two straight lines and a given point outside them a third straight line may be determined which passes through the point of intersection of the given lines and contains every point that is harmonically separated from the given point by the given lines.	From a straight line and two given points outside it a third point may be determined which lies on the straight line joining the given points, and through which passes every straight line that is harmonically separated from the given line by the given points.

In reality this proposition is merely a repetition of that of Art. 61, when the latter is transcribed for forms in the plane. Upon the theorem on the left and the one next to be stated rests the solution of the problem mentioned in the Introduction (Art. 6, see Fig. 1), viz.,

"Through the inaccessible point of intersection of two straight "lines to pass a third straight line."

64. After what has been said it will be an easy matter for you now to prove the following properties of the complete quadrangle and complete quadrilateral (see Fig. 15).

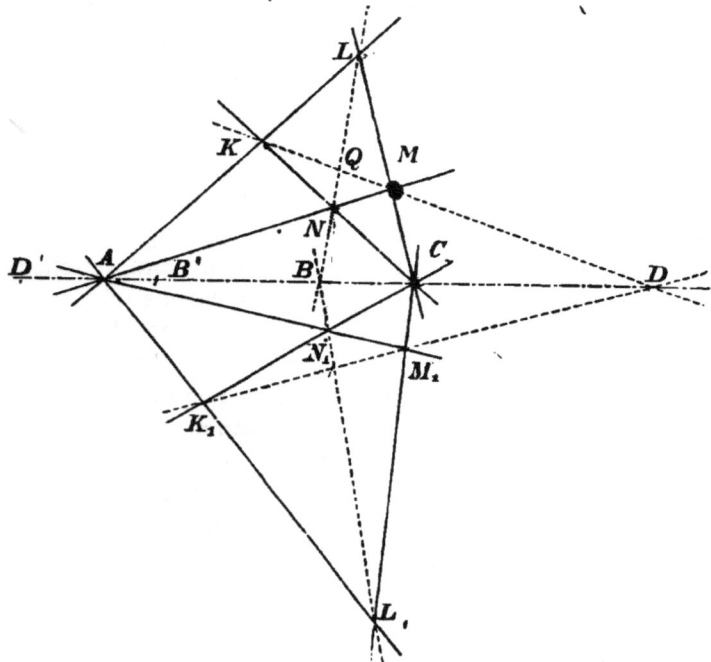

FIG. 15.

In the complete plane quadrangle any two opposite sides (as KM and LN) are harmonically separated by the two points A and C in which the remaining pairs of opposite sides intersect.

In the complete plane quadrilateral any two opposite vertices (as A and C) are harmonically separated by the two straight lines KM and LN which join the remaining pairs of opposite vertices.

You will notice that in Fig. 15 not only can K, L, M, and N be taken as vertices of a complete quadrangle, but also AL, AN, CL, and CN can be taken as sides of a complete quadrilateral of which A and C, K and M, L and N are the three pairs of opposite vertices.

65. In the quadrangle $KLMN$ (Fig. 15) let the two vertices
K, L, and the points of intersection A and C of two pairs
of opposite sides remain fixed, while the side MN rotates
about A, the vertices M and N moving along the straight lines
CL and CK, respectively; at the same time the two diagonals
will rotate about K and L, and the points B and D will so
move along the line AC that they remain harmonically separated
by A and C.   Since now no position of either point B or D
corresponds to more than one position of the other, it is impossible that these points should move at one time in the same
sense and at another time in opposite senses along AC; moreover,
they must always move in opposite senses since they are always
separated by A and C, and must coincide with C when the moving
line MN is brought into the position CA, or with A when brought
into the position LA.

From this it follows that—

If a pair of points A *and* C *harmonically separate each of two other
pairs of points* B, D *and* B_1, D_1, *then are* B *and* D *not separated by*
B_1 *and* D_1. *Two pairs of points on a straight line which mutually
separate each other cannot therefore both be harmonically separated
by the same third pair of points.*

66. To two pairs of points B, D, and B_1, D_1, on a straight line,
which do not separate each other, there always exists (at least)
one third pair of points A, C, by which B is harmonically separated
from D and at the same time B_1 from D_1.

In order to prove this, we shall imagine the segment $B_1 D_1$
upon which B and D do not lie to be traversed by a point P.
Of the points P_1 and P_2 which are harmonically separated from
P by B_1 and D_1, and by B and D, respectively, the first P_1
describes the supplement of the segment $B_1 D_1$ and the latter P_2
a segment $B_2 D_2$ contained in this supplement, whose extremities
are harmonically separated from B_1 and D_1, respectively, by B
and D.   The points P_1 and P_2 move in the sense opposite to
that in which P moves, and must coincide at least once, since P_1
describes a segment within which the segment traversed by P_2 is
contained.   If, now, we denote by A this point of coincidence and
by C the corresponding position of the point P, then A and C
harmonically separate B from D and at the same time B_1 from D_1.

METRIC RELATIONS OF HARMONIC FORMS.

67. I ought not to close the theory of harmonic forms without developing for you, as was suggested in the Introduction, their most important metric relations. We approach these most simply by means of the following theorem :

If in a straight line two points A *and* C *have equal distances from a third point* B, *then are they harmonically separated by this point and the infinitely distant point* D *of the straight line, or* A, B, C, D *form a harmonic quadruple.*

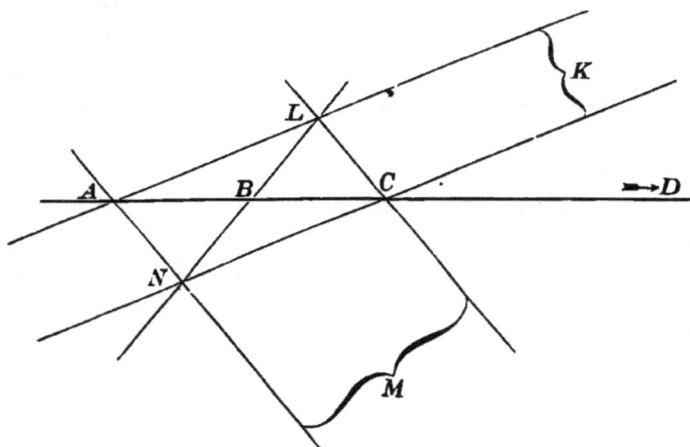

FIG. 17.

In a plane passed through the line *ABC* choose two infinitely distant points *K* and *M* (Fig. 17), and toward each of them draw parallel lines through *A* and *C*; these will intersect in two new points *L* and *N*. The straight line *LN*, as second diagonal of the parallelogram *ALCN*, passes through the bisection point *B* of the segment *AC*. Of the quadrangle *KLMN*, then, two opposite sides *KL* and *MN* intersect in *A*, two others *LM* and *NK* in *C*, the diagonal *LN* passes through *B*, and the second diagonal, namely, the infinitely distant straight line *KM*, passes through *D*; so that *ABCD* are in fact harmonic points.

68. Since four harmonic points *ABCD* are projected from a fifth point *S* by four harmonic rays, it follows that—

If we draw through the vertex *S* of a triangle *ASC*, two straight lines, the one *d* parallel to the base *AC*, and the other *b* toward the middle point of the base, then are these harmonically separated by the adjacent sides of the triangle.

If ASC is isosceles, then b is at right angles to AC, and consequently also to d; moreover, the supplementary angles formed by a and c are bisected by b and d. Hence:

"The lines bisecting two supplementary adjacent angles are "harmonically separated by the boundaries of these angles, and are "normal to each other."

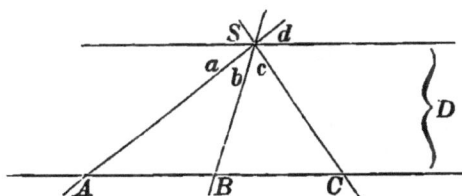

FIG. 18.

The converse of this theorem may be stated thus:

"If, of four harmonic rays, two conjugate rays are at right angles, "these bisect the angles between the other two."

The proof of this is derived immediately from the following statement, which will be recognized as the converse of the one just quoted:

"If a harmonic sheaf $abcd$ is cut by a straight line u parallel to "one of its rays, then one of the three points of intersection with the "remaining rays bisects the segment between the other two."

The points of intersection of u with $abcd$ are four harmonic points, and one of them lies infinitely distant.

69. These theorems, to which similar ones may be stated for harmonic planes, can be utilized for the solution of a series of problems. Thus, for example, the problem:

"To construct the fourth harmonic to three points or rays" admits a solution very much simpler than by means of the complete quadrangle as soon as the construction of parallels and of equal segments is conceded. For if to the rays b, c, d (Fig. 18), the fourth ray a, harmonically separated from c, is required to be found, we may intersect b and c by any straight line u parallel to d in the points B and C, and upon this line make AB equal to BC; the ray a of the sheaf bcd, passing through A, is the one sought.

If, further, to the three points A, B, C (Fig. 19), the fourth point D, harmonically separated from B, is required, we may lay off upon any straight line passing through B, two equal segments A_1B and BC_1; then determine the point of intersection S of the straight lines AA_1 or a and CC_1 or c, and draw through this point a line d parallel

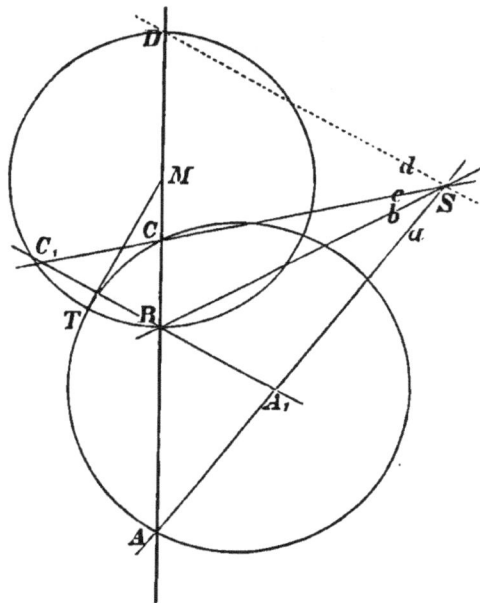

to A_1BC_1. This straight line will cut ABC in the required point D. For, since A_1, B, C_1, and the infinitely distant point of A_1B are four harmonic points, S (A_1BC_1D) or $abcd$ must be a harmonic sheaf, and therefore $ABCD$, a section of this sheaf, is a harmonic range of points.

70. If a segment AC and its middle point B are given, a parallel to the line ABC can be constructed through any point K (Fig. 20) with the use of the ruler only, as follows:

Draw the lines KA and KC and intersect them in L and N respectively by any straight line passed through B; then determine the point of intersection M of CL and AN, and through this point will -pass the required parallel. For, as second diagonal of the quadrangle $KLMN$, the line KM intersects the straight line ABC in a fourth point harmonically separated from B by A and C, but which lies infinitely distant since B bisects the segment AC.

If, conversely, two parallel lines are given, any segment lying upon one of them may be bisected by a linear construction. How these constructions can be turned to account, in land surveying for instance, will be evident to you without further comment.

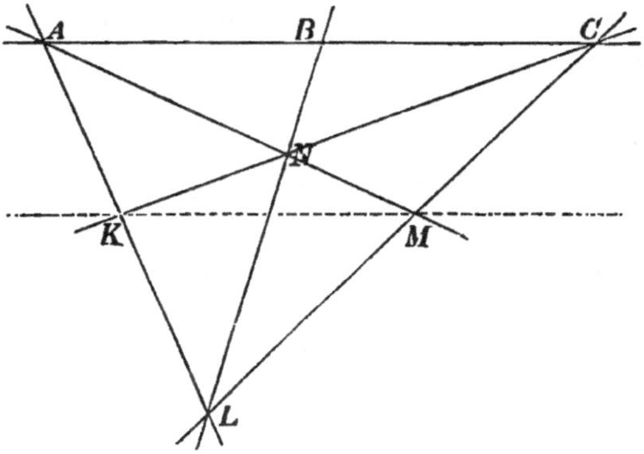

FIG. 20.

71. Among the segments which are formed by four harmonic points $ABCD$ on a straight line there exists an interesting proportion. In order to find this, we project the harmonic points from some centre S by a harmonic sheaf $abcd$ (Fig. 19) and then pass through B a parallel to the ray d. This meets the rays a and c in two points A_1 and C_1, which are equally distant from B; at the same time two pairs of similar triangles are formed, namely, AA_1B similar to ASD and BC_1C similar to DSC. We obtain then the proportions:

$$\frac{AB}{A_1B}=\frac{AD}{SD} \text{ and } \frac{BC}{BC_1}=\frac{CD}{SD}.$$

If we divide the first of these by the second, and take into consideration that $A_1B = BC_1$, we obtain

$$\frac{AB}{BC}=\frac{AD}{CD} \quad\dots\dots\dots\dots\dots\dots\dots\dots\dots\dots\dots(1)$$

The segment AC is divided internally at B in the same ratio as externally at D, B and D being harmonically separated by A and C.

This relation is frequently taken as the definition of harmonic points, and might be chosen as starting point in the study of the theory of harmonic forms.

As a consequence of this relation the point D lies out beyond C, if AB is greater than BC and likewise AD greater than CD; while, on the other hand, D lies out beyond A if AB is less than BC; so that both B and D are nearer to A than to C, or else both are nearer to C than to A.

72. In the above proportion, since the equal segments CD and DC are described in opposite senses, we ordinarily write $-DC$ instead of CD, so that the proportion reads more symmetrically, thus:

$$\frac{AB}{BC} = -\frac{AD}{DC}.$$

If M is the middle point of the segment BD, equation (1) may be written:

$$\frac{AM-BM}{BM-CM} = \frac{AM+MD}{CM+MD};$$

or if MD is replaced by BM:

$$\frac{AM-BM}{BM-CM} = \frac{AM+BM}{CM+BM}.$$

Clearing of fractions we obtain after a very simple reduction:

$$BM^2 = AM \cdot CM, \dots\dots\dots\dots(2)$$

or the remarkable property,

"BM (and similarly DM) is a mean proportional between "AM and CM."

This useful property might also be admitted as a definition of harmonic points.

73. If we draw any circle through A and C (Fig. 19), and to this a tangent MT from M, then by the well-known theorem upon the segments of secants of a circle,

$$AM \cdot CM = TM^2,$$

and hence $TM^2 = BM^2 = DM^2$. The point of contact T of the tangent lies therefore upon the circle which is described with centre M and radius BM or MD; and this circle cuts the other orthogonally at T, since its radius MT is tangent to the other at T. Thus,

"In a plane, the circles which pass through two given points "A and C are cut orthogonally by any circle, the extremities of "whose diameter BD are harmonically separated by A and C."

D

The theory of harmonic points thus leads us easily to the study of systems of circles which intersect orthogonally, and might suitably be chosen as the starting point in the investigation of orthogonal systems of spheres.

74. The reciprocal of equation (1), *i.e.* the equation $\dfrac{BC}{AB} = \dfrac{CD}{AD}$, may be written :

$$\frac{AC - AB}{AB} = \frac{AD - AC}{AD}, \text{ or } \frac{AB - AC}{AB} = \frac{AC - AD}{AD},$$

and it is to this latter equation that the points A, B, C, D, owe the name 'harmonic' points. For, as you are aware, we are accustomed to say of three quantities β, γ, δ that they are 'in harmonical progression,' or are 'harmonic,' if the difference between the first two is to the first as the difference between the last two is to the last, that is if

$$\frac{\beta - \gamma}{\beta} = \frac{\gamma - \delta}{\delta},$$

and the equation last written involving the segments AB, AC, and AD has exactly this form.

Performing the division indicated above we have finally the relation :

$$1 - \frac{AC}{AB} = \frac{AC}{AD} - 1,$$

which may be written in the form :

$$\frac{2}{AC} = \frac{1}{AB} + \frac{1}{AD}. \quad \dots\dots\dots\dots\dots\dots\dots(3)$$

You yourselves will easily be able to clothe this very remarkable formula in words ; it likewise is frequently used as the definition of harmonic points.

Similar equations might be developed for the angles formed by four harmonic rays or planes. I prefer, however, to present these incidentally in the supplement to the next main division of our subject, since they have no great value for us.

75. You will observe that the principle of reciprocity is not applicable to metric relations, or at least it applies only in individual cases. One reason for this is that in a sheaf there exists no element which occupies a distinctive position with reference to measurement similar to that of the infinitely distant point in the range of points ; while, on the other hand, we recognize in the latter no segment which could be so defined and characterized by measurement as can the right angle in the sheaf.

· EXAMPLES.

1. To three given elements in each of the one-dimensional primitive forms construct the respective fourth harmonic elements.

2. Through a given point draw a straight line which if produced would pass through the inaccessible point of intersection of two given lines.

3. Without the use of circles bisect a segment AC of a straight line, having given a parallel to this line.

4. In a plane are given a parallelogram and a segment AC of a straight line ; it is required, without the use of circles, to bisect AC and to draw a parallel to AC ; also to divide AC into n equal parts.

5. If A, B, C, D, are four harmonic points of a straight line, and a circle is described upon AC as diameter, of which S is any point, prove that the arc subtending the angle BSD, or its supplement, is bisected at A or at C.

6. Through a given point P draw a straight line meeting two given lines of the plane in A and B so that (1) the segment AB shall be bisected at P, (2) the segment AP shall be bisected at B. Under what circumstances is the solution impossible?

7. If two points are each harmonically separated from a third point by a pair of opposite edges of a tetrahedron, they are harmonically separated from each other by the third pair of opposite edges.

[The plane of the three points intersects the tetrahedron in a complete quadrilateral whose diagonals intersect in the three points.]

8. Given two pairs of points, A, B, and A_1, B_1, upon the same straight line, which do not separate each other. With the aid of circles, find two points which harmonically separate each pair.

[Choose any point D outside the given line and describe the circles DAB and DA_1B_1 intersecting a second time in E. Let the straight line DE cut the given line in O. From O draw a tangent OT to DAB or to DA_1B_1. The circle whose centre is O and radius OT will cut the given line in the required points.]

9. A straight line intersects the sides of a triangle ABC in the points A_1, B_1, C_1, and the harmonic conjugates A_2, B_2, C_2, of these points, with respect to the two vertices on the same side are determined, so that AC_1BC_2, BA_1CA_2, CB_1AB_2, are harmonic ranges. Show that A_1, B_2, C_2 ; B_1, C_2, A_2 ; C_1, A_2, B_2, are collinear, that AA_2, BB_2, CC_2, are concurrent, and that AA_2, BB_1, CC_1 ; AA_1, BB_2, CC_1 ; AA_1, BB_1, CC_2, are also concurrent.

LECTURE V.

76. In the present lecture I shall again take up and further
develop an idea which· has already been mentioned, namely, the
correlation of two primitive forms so that to each element of the
one there corresponds one and only one element of the other.
As very simple methods of correlating two primitive forms of the
first grade the following have already been mentioned :

(1) A sheaf of rays or planes and a range of points (Fig. 6), or
a sheaf of planes and a sheaf of rays, are correlated to each other
if each element of the latter lies upon the corresponding element
of the former.

(2) Two ranges of points are correlated to each other if they are
sections of one and the same sheaf of rays (Fig. 8).

(3) Two sheaves of rays are correlated to each other if they are
projectors of one and the same range of points (Fig. 7), or sections
of one and the same sheaf of planes, or both.

(4) Two sheaves of planes are correlated if they are projectors
of one and the same sheaf of rays.

77. We shall hereafter speak of two one-dimensional primitive
forms which are correlated to each other in any of the ways mentioned above, as being ' in perspective position,' or more briefly, they
shall be spoken of as being 'perspective' to each other ; so that
of two 'perspective' primitive forms of different kinds, the one is
always a section of the other, while on the other hand, two
'perspective' primitive forms of the same kind are always either
sections or projectors of one and the same third primitive form.

78. If two one-dimensional primitive forms are correlated perspectively to one and the same third form (for example, two ranges of points to a third), they are also correlated to each other, but in general are not in perspective position with regard to each other. We thus observe a second, the so-called 'skew' position of two correlated primitive forms, which might be obtained from the perspective position by giving one or the other of the perspective forms a slight displacement; then with each element of the one form ·there would continue to be associated a particular element of the other form, but in general the forms would lose their perspective position.

79. Two primitive forms may be correlated to each other, however, in numberless other ways; for example, two sheaves of rays are correlated by viewing them as projectors of one and the same curve, no matter what this curve may be. The method of correlation in question, however, is distinguished from all others in one important particular, and just as clearly so, whether the forms are in perspective or skew position; namely, if any four harmonic elements are selected from one of the two forms, to these evidently correspond four harmonic elements in the other form, since projectors and sections of harmonic forms are in turn harmonic forms. This peculiarity is not in general found in other methods of correlation and we are thus led to enunciate the following definition :

Two primitive forms are said to be related 'projectively' to each other, or, more briefly, are said to be 'projective,' when they are so correlated that to every set of four harmonic elements in the one form there correspond four harmonic elements in the other.

Two perspective one-dimensional primitive forms are therefore also projective, perspective correlation and projective correlation being distinguished only by the relative positions of the forms. The expressions 'conformal' and 'homographic,' which Paulus and Chasles use, have the same meaning as 'projective.' Von Staudt has introduced the symbol $\overline{\wedge}$ for 'is projective to.'

From the definition of the projective relation it follows immediately that—

If two forms are each projective to a third form they are projective to each other.

For example, if two ranges of points are perspective, and hence projective, to one and the same third range, then are they projective to each other, but it is only in particular cases that they

are in perspective position relative to each other. The same is.
true of any two primitive forms of the first grade.

80. *In two projective ranges of points, to any four points* A, B, C, D,
*of the one range, of which the first two are not separated by the last
two, there correspond always in the other range four points* A_1, B_1,
C_1, D_1, *which are subject to the same condition.*

For there are, in the first range, two points M and N by which
A is harmonically separated from B, and also C from D (Art. 66),
and in accordance with the definition of projectivity there correspond to these, in the other range, two points M_1 and N_1 by which
A_1 is harmonically separated from B_1, and also C_1 from D_1. On
this account it is impossible for the points A_1 and B_1 to be
separated from each other by C_1 and D_1 (Art. 65). If A and C are
separated by B and D, then must also A_1 and C_1 be separated by
B_1 and D_1, for the opposite conclusion would bring us into dis-
agreement with what has just been proved.　　　　-

If in the one range any number of points A, B, C, ... P, Q, ...
are so chosen that no two of them are separated by the one named
just before and the one just after them, there correspond to these, in
the other range, the same number of points A_1, B_1, C_1, ... P_1, Q_1, ...
for which the same relation holds true. If the points P, Q, R, ... of
the first range are consecutive points of the range, then must also their
corresponding points P_1, Q_1, R_1, ... be consecutive points in the second
range, for if P_1 and Q_1, say, were not consecutive points of this range,
there would be points U_1, V_1, which separate them, and it would
be necessary then that P and Q be separated from each other by
the corresponding points U and V, and they could not in that
case be consecutive points. Similar conclusions may be reached
in the case of sheaves of rays and of planes, since these are cut
by arbitrary transversals in projective ranges of points. Hence the
important theorem :

*If two one-dimensional primitive forms are projectively related,
then to every continuous succession of elements of the one form there
corresponds a continuous succession of elements of the other.*

81. Two projective primitive forms of the same kind may also
be *conjective* or be *superposed*, that is to say, may have the same
base. Two projective sheaves of planes, for example, may be placed
with their axes coinciding, and similarly two projective ranges of
points may lie on the same straight line, so that each point of the
line must be considered twice, once as belonging to the one range,

and again as belonging to the other range. The investigation as to how many 'self-corresponding' elements may exist in two projective one-dimensional primitive forms which are superposed, that is, how many elements of one form coincide with their homologous elements in the other, is of great importance in all that follows.

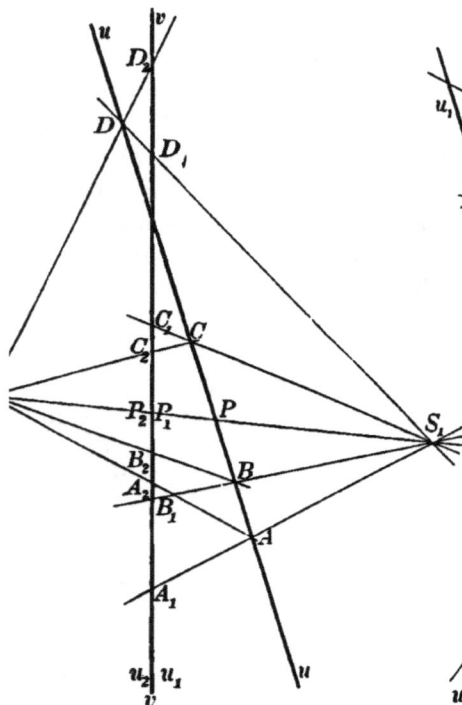

FIG. 21.

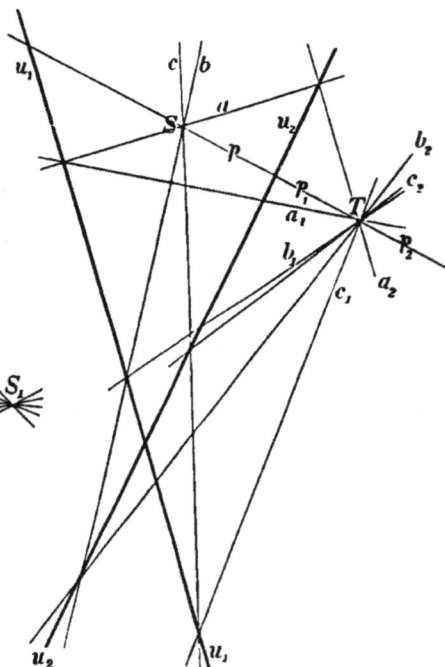

FIG. 22.

82. In the first place, that it is possible for the two forms to have either one or two such 'self-corresponding elements' is clear from the following theorem:

If in a plane there are given two sheaves of rays S_1 and S_2 (Fig. 21) which are projectors of one and the same range of points μ, *i.e.* are perspective, and these are intersected by a straight line v, then there are determined upon this straight line two projective

If in a plane there are given two ranges of points u_1 and u_2 (Fig. 22) which are sections of one and the same sheaf S, *i.e.* are perspective, and these are projected from a point T of the plane, then this point becomes the centre of two projective sheaves of rays in which ranges of points u_1 and u_2, in which the intersection of v with u and with $S_1 S_2$ are self-corresponding points. These two points coincide if $S_1 S_2$ passes through uv.

the lines joining T with S and with $u_1 u_2$ are self-corresponding rays. These two rays coincide if $u_1 u_2$ lies upon ST.

In the theorem on the left two points A_1 and A_2 of u_1 and u_2, respectively, correspond to each other if $S_1 A_1$ and $S_2 A_2$ intersect in a point A of u.

The three ranges of points u, u_1 and u_2 therefore have the intersection point uv self-corresponding, while u_1 and u_2 have also self-corresponding the intersection point of v with the common ray $S_1 S_2$.

83. If, of two projective ranges of points u and u_1, the first is described by the continuous motion of a point P, the corresponding point P_1 at the same time moves continuously along the other

FIG. 23.

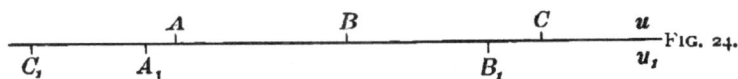

FIG. 24.

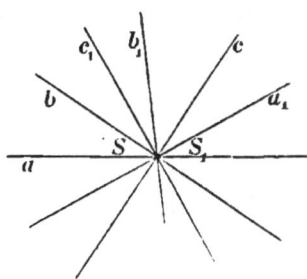

FIG. 25.

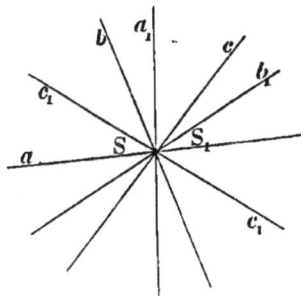

FIG. 26.

range, and if u and u_1 lie upon the same straight line, the points P and P_1 may move either in the same sense or in opposite senses (Figs. 23 and 24) upon the line. In the case of figure 24, in which the points move in the same sense along the line, we shall call the ranges *directly* projective, in the other case (Fig. 23) *oppositely* projective. We shall in the same way call two projective sheaves of rays S and S_1 which are concentric and lie in the same plane (Figs. 25 and 26), or two projective sheaves of planes which have the same axis, 'directly' projective (Fig. 26), or 'oppositely projective' (Fig. 25), according as two homologous elements in describing the sheaves would rotate in the same or in opposite senses. In primitive forms of the first grade, which are superposed and are oppositely projective, the two describing elements must necessarily coincide twice; hence,

"Primitive forms of the first grade which are superposed and are "oppositely projective, always have two self-corresponding elements; "all other pairs of homologous elements are separated by these "self-corresponding elements."

On the other hand, forms which are superposed and are directly projective have two self-corresponding elements only in case a segment AB (or an angle) of the one lies wholly within the corresponding segment (or corresponding angle) of the other (Fig. 24); in particular cases they have only one, and may have (Fig. 26) no self-corresponding element. Two projective ranges of points u and u_1 which have three self-corresponding points A, B, C, must be directly projective.

84. We can now prove the following fundamental theorem of the Geometry of Position:

If two projective one-dimensional primitive forms have three self-corresponding elements A, B, C, *then are all their elements self-corresponding and the forms are consequently identical.*

Suppose, in the first place, that the projective primitive forms are two ranges of points u and u_1 (Fig. 27). Then a point which

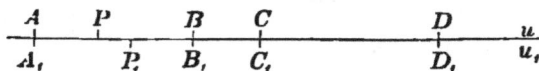

FIG. 27.

is harmonically separated from any one of the self-corresponding points A, B, C, by the other two must coincide with its corresponding point, since four harmonic points of u always correspond to four harmonic points of u_1 by definition, and the harmonic conjugate of a point with respect to two others is uniquely determined. Suppose further that there is given upon that segment AB which does not contain C, a point P of u which does not coincide with its corresponding point P_1 of u_1. If now we permit P to traverse the range u in the sense ABC, then P_1 will traverse u_1 in the same sense, and must coincide with P in B, or perhaps before reaching B in some point B_1. If P should move in the opposite sense CBA, then P_1 would move in the sense CBA, and would coincide with P in A, or before reaching A in a point A_1. In this way we should obtain a segment A_1B_1 which is either equal to AB or is a part of AB, and of which no point, except the extremities A_1 and B_1, coincides with its corresponding point. But this is impossible, since that point which is harmonically separated from C by A_1 and B_1 must coincide with its corresponding point and fall in this segment. Hence the ranges u and u_1 must have every point of the segment AB self-corresponding, and therefore every other point $Q_.$ self-corresponding, since Q is harmonically separated from some point of the segment AB by A and B.

The theorem may be proved in an analogous manner for the case of two projective sheaves of rays or planes which have three self-corresponding elements, or what is simpler, these cases may be reduced to the one just proved by intersecting the sheaves with a straight line. The ranges of points thus obtained would be projective and have three self-corresponding elements; consequently all their elements would be self-corresponding; hence all the elements of the sheaves must be self-corresponding.

85. Two projective one-dimensional primitive forms can have then at most two self-corresponding elements, unless every element of the one coincides with the corresponding element of the other. An important deduction from this fundamental theorem is the following:

"If a range of points is projective to any sheaf, or a sheaf of "rays to a sheaf of planes, and three elements of the first form lie "upon their corresponding elements in the second form, then the "first form is a section of the second."

For it has three self-corresponding elements, with that section of the second form which is made by its base; hence all their elements are self-corresponding, and the first form is identical with the section of the second form.

86. If two projective sheaves of rays S and S_1 (Fig. 29) lying in the same plane, but not concentric, have their common ray a (or a_1) self-corresponding, then are they projectors of one and the same range of points u, and are consequently in perspective position.

If two projective ranges of points u and u_1 (Fig. 28) which lie in the same plane, but are not conjective, have their point of intersection A (or A_1) self-corresponding, then are they sections of one and the same sheaf of rays S, and are consequently in perspective position.

For if we join the points B and C, in which any two rays b and c of the sheaf S are intersected by their corresponding rays b_1 and c_1 of S_1, by a straight line u, then are the two ranges of points in which u cuts the sheaves S and S_1 identical, since they are projectively related, and have three self-corresponding points ua, ub, and uc.

For if we join any two points B and C of u with their corresponding points B_1 and C_1 of u_1 by the straight lines b and c, and denote the point of intersection of these lines by S, then are the two sheaves of rays by which the ranges of points u and u_1 are projected from S_1 identical, since they are projective, and have three self-corresponding rays SA, SB, and SC.

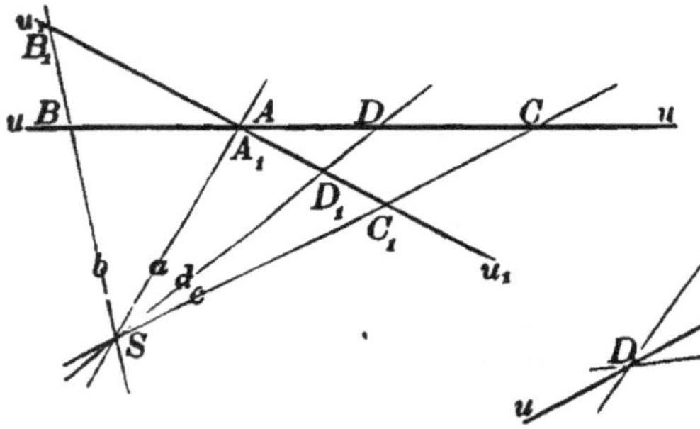

FIG. 28. 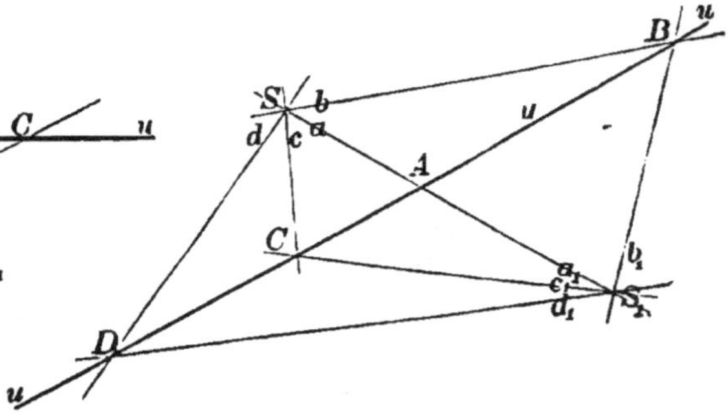　　　　　FIG. 29.

87. The following theorems from the geometry of the bundle of rays are analogous to those just stated, and may be proved in a similar manner:

If two projective sheaves of planes whose axes intersect have the plane of the axes self-corresponding, then are they projectors of one and the same sheaf of rays, and are consequently perspective.

For if we intersect the two sheaves of planes with a plane which is determined by the lines of intersection of any two pairs of homologous planes of the sheaves, we obtain two projective sheaves of rays which have three self-corresponding rays, and which are consequently identical.

If two projective sheaves of rays which are concentric, and lie in different planes have the line of intersection of these planes self-corresponding, then are they sections of one and the same sheaf of planes, and are therefore perspective.

For if we project the two sheaves of rays from the line of intersection of any two planes, each of which is determined by a pair of homologous rays of the sheaves, we obtain two projective sheaves of planes which have three self-corresponding planes, and which are consequently identical.

88. Two projective but not con-centric sheaves of rays S and S_1 (Fig. 29) have perspective position if, of the points of intersection of pairs of homologous rays, any three, B, C, and D, lie in one straight line u.

For, the projective ranges of points in which the two sheaves of rays are cut by the straight line u have these three points B, C, D, and consequently all their points self-corresponding ; all points of intersection of homologous rays of the two sheaves consequently lie upon the straight line u.

Two projective but not conjective ranges of points u and u_1 (Fig. 28) have perspective position if, of the lines joining their homologous points, any three, BB_1, CC_1, and DD_1, pass through one point S.

For, the projective sheaves of rays by which the two ranges of points are projected from the point S have these three rays BB_1, CC_1, DD_1, and consequently all their rays self-corresponding ; all lines joining homologous points of u and u_1 consequently pass through S.

You can easily enunciate and prove the analogous theorems for sheaves of rays and planes in the bundle of rays.

89. If two correlated sheaves of rays lie in one plane and are not concentric, the points of intersection of pairs of homologous rays form a continuous curve, for if a ray describes the one sheaf by continuous rotation about its centre, the corresponding ray will rotate continuously about the other centre, and will describe the other sheaf; the point of intersection of the two rays will conse-quently describe a continuous line.

If now the two sheaves of rays are projectively related but are not in perspective position, all the points of intersection of pairs of homologous rays lie upon a curve which, in consequence of the theorem of Art. 88, has in common with no straight line more than two points. On account of this peculiarity we shall call this curve a *curve or range of points of the second order*; the ordinary range of points shall be distinguished from this, wherever it appears necessary, by the name *range of points of the first order*.

If two projective ranges of points lie in the same plane and are not perspective, then, as in the case above, the straight lines joining pairs of homologous points form a continuous succession of rays, of which not more than two pass through any point of the plane. We designate the totality of these joining lines as a *sheaf of rays of the second order*, the ordinary sheaf of rays being hereafter de-signated a *sheaf of the first order* to avoid confusion.

Curves and sheaves of rays of the second order are defined from the nature of their formation, then, in the following manner:

Two projective sheaves of rays (of the first order), which lie in the same plane and are neither concentric nor perspective, 'generate' a curve or range of points of the second order, each ray of the one sheaf intersecting the corresponding ray of the other in a point of this curve.

Not more than two points of this curve of the second order lie upon any straight line.

Two projective ranges of points (of the first order), which lie in a plane and are neither superposed nor perspective, 'generate' a sheaf of rays of the second order, each point of the one range being projected from the corresponding point of the other by a ray of this sheaf.

Not more than two rays of a sheaf of the second order pass through any point.

Wholly analogous forms of the second order are generated in a bundle by projective sheaves of rays and planes. All these new forms will be investigated more closely in the next lecture.

90. The construction of curves and sheaves of rays of the second order depends upon the following important theorem:

Two one-dimensional primitive forms may always be correlated projectively to each other so that any three elements of the one shall correspond to three elements of the other chosen at random; to any fourth element of the one form, the corresponding element of the other is then uniquely determined.

The proof of this theorem might be deduced directly from the definition of projectivity and the fact that, by three elements of a one-dimensional primitive form, a *single* fourth element is determined, which is harmonically separated from one of these three by the other two. From considerations similar to those of Article 84, it appears that by means of the three given pairs of homologous elements indefinitely many such fourth pairs are correlated to each other, and that no element of the one form is without a corresponding element in the other.

I shall, however, proceed to prove the theorem in a different manner, but only for two ranges of points, since all other cases may be easily reduced to this one. For instance, if one or each of the two primitive forms is a sheaf, then instead of this we can substitute its section by a straight line, which is a range of points.

91. Suppose, in the first place, that two ranges of points u and u_1 lying in one plane (Fig. 28), are so correlated projectively that they

have the point of intersection A (or A_1) of their bases self-corresponding, and that the points B and C of u correspond, respectively,
to the points B_1 and C_1 of u_1; then must the one range of points
be a projection of the other from S, the point of intersection of
BB_1 and CC_1. To any fourth point D of u corresponds, then,
in u_1 its projection D_1 from the point S.

Next, suppose that any two ranges of points u and u_1 (Fig. 30),
not lying on the same straight line, are so related projectively that

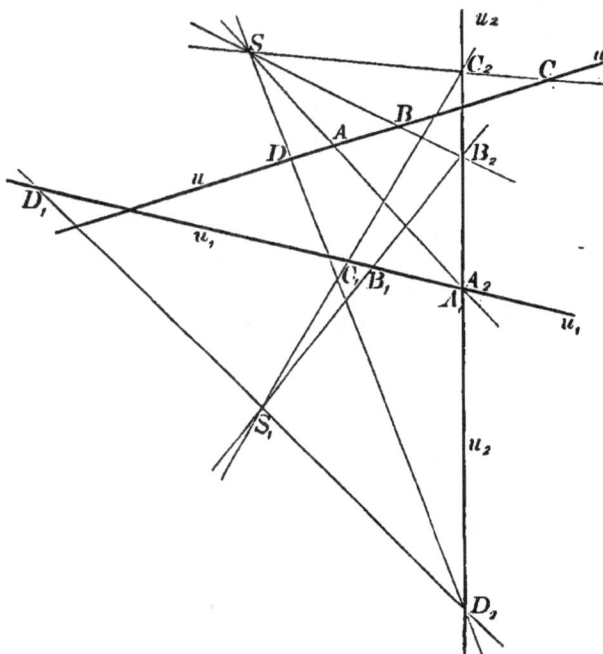

FIG. 30.

the points A, B, C, of u correspond respectively to the points
A_1, B_1, C_1, of u_1. Let us now choose upon one of the straight
lines joining two corresponding points, for example upon AA_1, a
point S different from either A or A_1, and draw through A_1 a
straight line u_2 cutting u, and different from either u_1 or AA_1.
Project now the range u upon u_2 from the centre S, and let
A_2, B_2, C_2, be the projections of A, B, C, upon u_2. By this means
our problem is reduced to that of the preceding paragraph. For
we have only to so correlate u_1 and u_2 that they may have their point
of intersection A_1 (or A_2) self-corresponding, and the points B_1

and C_1 of u_1 corresponding respectively to B_2 and C_2 of u_2. The ranges of points u and u_1 can then be looked upon as projections of one and the same third range of points u_2. In order to determine the corresponding point D_1 in u_1 to any point D in u, we find first its projection D_2 upon u_2, by the aid of which we can then find D_1 according to the method already pointed out.

Suppose, finally, two ranges of points lying upon the same straight line are so correlated projectively to each other that the points A, B, C, of u correspond respectively to the points A_1, B_1, C_1, of u_1. This case can be reduced to the preceding by first projecting u_1 upon another straight line u_2. If any two corresponding points coincide, for example, A and A_1, u_2 is drawn most advantageously through that point, and so the problem is referred back to the first case considered.

92. During the process of these investigations there has arisen the following theorem :

Two projective one-dimensional primitive forms may always be considered as the first and last of a series of forms, of which each is perspective both to the one preceding it and to the one following it.

Two projective ranges of points, for example, may be looked upon as the first and last of a series of not more than four ranges of points, of which each is a projection of the adjacent ones. This fact justifies the use of the term 'projective.'

93. At the same time we obtain very simply a series of apparently complicated theorems ; of these I shall mention only the following :

If the sides a_1, a_2, ... a_n, of a variable simple n-point rotate in order about n fixed points S_1, S_2, ... S_n, while $n-1$ vertices a_1a_2, a_2a_3, ... $a_{n-1}a_n$, of the same move along fixed straight lines u_1, u_2, ... u_{n-1}, respectively, then the remaining vertex, and likewise every other point of intersection of sides of the n-point, describes either a curve of the second order or a straight line ; a straight line certainly in case the fixed centres of rotation S_1, S_2, ... S_n, all lie upon one fixed straight line.

If the vertices A_1, A_2, A_3, ... A_n, of a variable simple n-side move in order along n fixed straight lines u_1, u_2, ... u_n, while $n-1$ of its sides A_1A_2, A_2A_3, ... $A_{n-1}A_n$, rotate about the fixed points S_1, S_2, ... S_{n-1}, respectively, then the remaining side A_nA_1, and likewise each diagonal of the n-side, either describes a sheaf of rays of the second order or else it rotates about a fixed point ; the latter case certainly happens if the straight lines u_1, u_2, ... u_n, all intersect in one fixed point.

The sides a_1, a_2, ... a_n, describe about S_1, S_2, ... S_n, sheaves of rays of which each lies perspective to the following one, the ranges of points u_1, u_2, ... u_{n-1}, forming perspective sections of these sheaves; consequently, all pairs of these sheaves, and in particular the first and last, are projectively related, and generate a curve of the second order if they do not chance to be perspective. This latter case happens, among other ways, if the centres of the sheaves all lie upon one straight line ; for in that case this line would be a self-corresponding ray of all the sheaves, since in the variation of the n-point all the sides at one time coincide with the line of centres.

The vertices A_1, A_2, ... A_n, describe upon u_1, u_2, ... u_n, ranges of points, of which each lies perspective to the following one, while S_1, S_2, ... S_{n-1}, form the respective centres of projection ; consequently, all pairs of these ranges of points, and in particular the first and last, are projectively related and generate a sheaf of the second order if they do not chance to be perspective. This latter case happens, among other ways, if the straight lines u_1, u_2, ... u_n, intersect in one point P, for in that case the ranges all have this point self-corresponding, since in the variation of the n-side all the vertices at one time coincide with P.

These theorems, of which the one on the left is a generalization of a theorem by Maclaurin and Braikenridge, and the other is due to Poncelet, afford us a means of finding any desired number of points of a curve or rays of a sheaf of the second order by linear constructions. These constructions are simplest when $n = 3$.

METRIC RELATIONS OF PROJECTIVE PRIMITIVE FORMS OF THE FIRST GRADE.

94. Among the angles and segments which are formed by any four homologous elements in two projective primitive forms there exists an important proportion, which in conclusion I shall now develop. Let us set out with a sheaf of rays S and a range of points u perspective to it (Figs. 28 and 29). Any four rays a, b, c, d, of S pass through their corresponding points A, B, C, D, of u. The triangles formed by u and two of these rays, whose common vertex is S, have equal altitudes ; their areas therefore vary as their bases, so that, for example :

$$\frac{\triangle ASB}{\triangle ASD} = \frac{AB}{AD} \text{ and } \frac{\triangle CSB}{\triangle CSD} = \frac{CB}{CD}.$$

But the area of a triangle is equal to half the product of two sides and the sine of the included angle. If we denote the angle between two rays p and q by (pq), we obtain for the areas of the four triangles the values :

$$\triangle ASB = \tfrac{1}{2}AS . SB . \sin(ab) ;$$
$$\triangle ASD = \tfrac{1}{2}AS . SD . \sin(ad) ;$$
$$\triangle CSB = \tfrac{1}{2}CS . SB . \sin(cb) ;$$
$$\triangle CSD = \tfrac{1}{2}CS . SD . \sin(cd) ;$$

and if these values are substituted in the above equations, and the common factors in numerator and denominator are removed, it follows that

$$\frac{SB . \sin(ab)}{SD . \sin(ad)} = \frac{AB}{AD} \quad \text{and} \quad \frac{SB . \sin(cb)}{SD . \sin(cd)} = \frac{CB}{CD}.$$

Dividing the first of these equations by the second we obtain :

$$\frac{\sin(ab)}{\sin(ad)} : \frac{\sin(cb)}{\sin(cd)} = \frac{AB}{AD} : \frac{CB}{CD}. \quad \dots\dots\dots\dots(1)$$

95. Each term of this proportion is a ratio ; for example, $\dfrac{CB}{CD}$ is the ratio of the two segments into which BD is divided by the point C (which in Fig. 28 lies outside BD), and $\dfrac{\sin(cb)}{\sin(cd)}$ is the ratio of the sines of the two angles into which the angle (bd) is divided by the ray c. Thus the left-hand side, and similarly the right-hand side, of this equation is a ratio between two ratios, or a so-called double-ratio, - cross-ratio, or anharmonic ratio. You will readily observe the particular way in which these anharmonic ratios are formed. That on the right, for example, among the segments formed by A, B, C, D, is obtained by dividing the ratio between the segments of BD determined by one of the remaining points A, by the ratio between the segments determined by the other point C, these segments being taken in the same order in both ratios. The anharmonic ratio among the sines of the angles is formed in an entirely analogous manner. Moreover, it follows from the manner of deriving our equation that it is immaterial which segment and which angle we look upon as being divided, if only both anharmonic ratios are formed in the same way. A different choice of our four triangles would have yielded the following relation :

$$\frac{\sin(ad)}{\sin(ac)} : \frac{\sin(bd)}{\sin(bc)} = \frac{AD}{AC} : \frac{BD}{BC},$$

E

and it would not be difficult for you to produce still other similar relations for the same points and rays.

96. It is worthy of note that this relation remains true if we alter the position of u relative to the sheaf of rays. Hence upon any other straight line u (Fig. 28) which cuts the rays a, b, c, d, in points A_1, B_1, C_1, D_1, respectively, segments are formed for which the following relation, analogous to (1), is true :

$$\frac{\sin(ab)}{\sin(ad)} : \frac{\sin(cb)}{\sin(cd)} = \frac{A_1B_1}{A_1D_1} : \frac{C_1B_1}{C_1D_1} ; \quad \dots\dots\dots\dots(2)$$

and similarly if A, B, C, D, are projected from any point S_1 (Fig. 29) different from S, by four rays a_1, b_1, c_1, d_1, then,

$$\frac{\sin(a_1b_1)}{\sin(a_1d_1)} : \frac{\sin(c_1b_1)}{\sin(c_1d_1)} = \frac{AB}{AD} : \frac{CB}{CD} \quad \dots\dots\dots\dots(3)$$

An anharmonic ratio among four elements of a range of points or of a sheaf of rays of the first order does not therefore alter its value if these elements are replaced by the corresponding elements of a perspective range or sheaf. (Compare Pappus, *Mathematicae Collectiones*, VII. 129.)

97. Since now we can always consider two one-dimensional primitive forms, which are projectively related, as the first and last of a series of primitive forms of which each is perspective to those adjacent, it follows that—

If two primitive forms are projective, then every anharmonic ratio among four elements of one of them is equal to the analogous anharmonic ratio among the four corresponding elements of the other.

If, for example, u and u_1 are two projective ranges of points, and to the points A, B, C, D, of u correspond the points A_1, B_1, C_1, D_1, of u_1, then there exists among the segments which are formed by these points the proportion :

$$\frac{AB}{AD} : \frac{CB}{CD} = \frac{A_1B_1}{A_1D_1} : \frac{C_1B_1}{C_1D_1}. \quad \dots\dots\dots\dots(4)$$

Steiner used this property as definition of the projective relation, and in this way placed the treatment of anharmonic ratios at the foundation of his *Systematische Entwickelung*, etc. The theorem holds equally well for sheaves of planes, since the angles between the planes of such a sheaf are measured in a sheaf of rays whose plane is at right angles to the axis of the sheaf of planes, and which is perspective to the latter.

98. In conclusion, it would be well to state here the metric properties of the angles of a harmonic sheaf of rays or planes. If a, b, c, d, are four elements of such a sheaf, and A, B, C, D (Fig. 19), the four harmonic points in which these are cut by a straight line, we have for the latter the relation $\dfrac{AB}{BC} = -\dfrac{AD}{DC}$ (Art. 72), and therefore from equation (1):

$$\frac{\sin(ab)}{\sin(bc)} = -\frac{\sin(ad)}{\sin(dc)}. \quad \ldots\ldots\ldots\ldots\ldots\ldots(5)$$

EXAMPLES.

1. (*a*) Two ranges of points u and u_1 are each perspective to a third range u_2; construct the sheaf of rays generated by u and u_1. When is this of the first and when of the second order?

(*b*) Two sheaves of rays S and S_1 are each perspective to a third sheaf S_2; construct the range of points (curve) generated by S and S_1. When is this of the first and when of the second order?

2. (*a*) Of two projective ranges of points u and u_1, three pairs of homologous points are given; to any fourth point D of u construct the corresponding point D_1 of u_1, whether u and u_1 lie upon the same or upon different straight lines.

(*b*) State and perform the reciprocal operation to (*a*).

3. To bring two projective sheaves of rays or ranges of points into perspective position.

4. To bring a range of points and a sheaf of rays projective to it into perspective position.

5. In one of two given perspective sheaves of rays find two rays normal to each other which correspond respectively to two normal rays in the other sheaf, and hence show that in two projective sheaves of rays whose centres do not lie infinitely distant there is always a pair of homologous right angles.

6. If a sheaf of planes u is perspective to a sheaf of rays S, the axis of the sheaf of planes is normal to one of the two rays at right angles in S which correspond to two planes at right angles in u. Thus we find that each of the two following problems admits of two solutions:

Given a sheaf of rays S and a sheaf of planes u projective to it, it is required—

(1) to pass a plane through a given point which shall intersect u in a sheaf of rays congruent to S.

(2) to find an axis from which S is projected by a sheaf of planes congruent to u.

7. Place a sheaf of rays S and a sheaf of planes u in such relative positions that three given planes α, β, γ, of u shall pass through three given rays a, b, c, of S (problem 6).

8. Construct a plane cutting the lateral faces α, β, γ, of a triangular prism in a triangle abc which is similar to a given triangle $a_1b_1c_1$. [This problem can be reduced to the preceding.]

9. Given two fixed straight lines u and u' intersecting in O, and two points S and S' collinear with O. A straight line v rotates about a fixed point U and intersects u and u' in A and A', respectively. Show that the locus of the point of intersection of the straight lines SA and $S'A'$ is a straight line. [Chasles, *Géométrie Supérieure*, Paris, 1880, Art. 342. Pappus, *Math. Coll.*, Book VII., Props. 138, 139, 141, 143.]
State also and prove the reciprocal theorem.

10. P, S, and S' are three collinear points, u and u' two fixed lines which intersect in O. Through P an arbitrary straight line v is drawn which intersects u and u' in A and A', respectively; the straight lines AS and $A'S'$ intersect in M. Show that as v rotates about P, M will move upon a straight line which passes through O. [Chasles, *loc. cit.*, Art. 343.]
This proposition may also be stated as follows:
If the three sides of a variable triangle MAA' rotate about three fixed collinear points P, S', S, respectively, while two vertices A and A' move upon two fixed straight lines which intersect in O, then the third vertex M will describe a straight line which also passes through O.
In this form the proposition is equivalent to Desargues' theorem upon perspective triangles stated in Article 7. State the reciprocal theorem.

11. If the four vertices A, B, C, D, of a variable quadrangle move respectively upon four fixed straight lines which pass through one point O, while three of the sides AB, BC, CD, rotate about three fixed collinear points, then the remaining three sides will also rotate about fixed points, these six fixed points forming the vertices of a complete quadrilateral, *i.e.* they lie three by three upon four straight lines. [Cremona, *Projective Geometry*, Oxford, 1885, Art. 111.]

LECTURE VI.

CURVES, SHEAVES, AND CONES OF THE SECOND ORDER.

99. In the last lecture we reached the following important results :

If two projective sheaves of rays lie in one plane, but are neither concentric nor perspective, the points of intersection of their homologous rays form a curve or range of points of the second order, which has not more than two points in common with any straight line.

If two projective ranges of points lie in one plane, but are neither upon the same straight line nor perspective, the lines joining pairs of homologous points form a sheaf of rays of the second order, which has not more than two rays in common with any sheaf of the first order.

In order to give you a definite conception of these forms of the second order, I shall state now, giving proofs later, that the curve of the second order is identical with the conic section, and hence may be obtained by intersecting an ordinary circular cone with a plane. A sheaf of rays of the second order consists of the system of tangents to such a conic section.

100. To the two preceding theorems from plane geometry, there correspond the following from the geometry of the bundle of rays :

If two projective sheaves of planes whose axes intersect are not perspective, the lines of intersection of their homologous planes form a cone of the second order, which has not more than two rays in common with any plane. The point of intersection of the two axes, through which all rays of the

If two projective sheaves of rays whose planes intersect are concentric but not perspective, the planes determined by pairs of homologous rays form a sheaf of planes of the second order, which has not more than two elements in common with any sheaf of planes of the first order. The

cone pass, is called the 'vertex' of the cone.

common centre of the sheaves of rays, through which all planes of the generated sheaf pass, is called the 'vertex' of the sheaf of planes.

The cone and, the sheaf of planes of the second order may be derived from the curve and the sheaf of rays of the second order by projecting the latter forms from some point not lying in their plane. For, the two projective sheaves of rays S and S_1 (Figs. 31 and 33), by which a curve of the second order is generated, are projected from such a point O (your eye, for example) by two projective sheaves of planes which generate a cone of the second order having its vertex at O and passing through the given curve. In the same way the two projective ranges of points u and u_1 (Fig. 32) which generate a sheaf of rays of the second order are projected from O by two projective sheaves of rays, and these in turn generate a sheaf of planes of the second order which passes through the given sheaf of rays and has O for vertex.

Conversely, every cone of the second order is intersected by a plane not passing through its vertex in a curve of the second order; for, the two projective sheaves of planes which generate the cone are intersected in two projective sheaves of rays which generate the curve of section. If, then, more than two rays of the cone were to lie in one plane, more than two points of the curve of section would lie in one straight line, which, from the theorem quoted at the opening of the lecture, is impossible. You will easily be able to prove for yourselves the analogous property of a sheaf of planes of the second order, and at the same time you will recognize the correctness of the following statements:

Any curve or any sheaf of rays of the second order is projected from a point not lying in its plane by a cone or by a sheaf of planes of the second order.

Any cone or any sheaf of planes of the second order is intersected by a plane not passing through its vertex in a curve or in a sheaf of rays of the second order.

101. You will observe from this that all results which are obtained for plane forms of the second order may be immediately carried over by projection to the analogous forms in the bundle of rays. I shall confine myself, therefore, in the first place to the investigation of curves and sheaves of rays of the second order, and shall begin with the following observation:

The curve of the second order k^2, which is generated by two projective sheaves of rays S and S_1 (*Fig.* 31), passes through the centres of these sheaves.

For to the ray SS_1, or p, of the sheaf S, *i.e.* to the line joining the two centres, corresponds in the sheaf S_1 a ray p_1 different from S_1S, since the sheaves are not perspective ; the point of intersection of p and p_1, namely S_1, lies therefore upon the curve k^2, and similarly it may be shown that the centre S is also a point of the curve.

The sheaf of rays of the second order K^2, which is generated by two projective ranges of points u and u_1 (*Fig.* 32), contains the straight lines u and u_1 upon which the ranges of points lie.

For to the point uu_1, or P, of the range u, *i.e.* to the point of intersection of the two straight lines, corresponds in u_1 a point P_1 different from u_1u, since the ranges of points are not perspective ; the joining line PP_1 or u_1 belongs therefore to the sheaf K^2, and similarly it may be shown that u ikewise belongs to the sheaf.

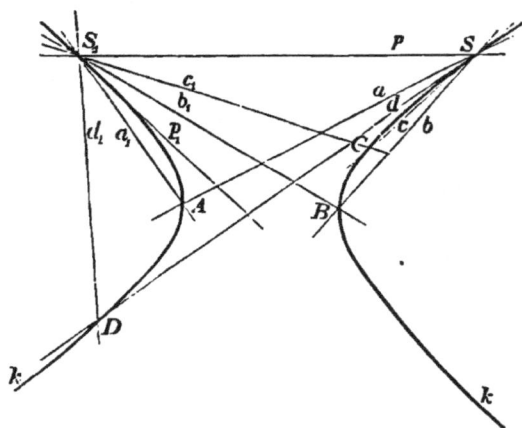

FIG. 31.

102. In the theorem on the left, of the preceding article, the ray p_1 (Fig. 31) is the only ray passing through S_1 which has but one point, namely S_1, in common with the curve. Any ray a_1 of the sheaf S_1, different from p_1, is intersected by its corresponding ray a, in a second point of the curve k^2, not coinciding with S_1. We say therefore that the ray p_1 *touches* the curve k^2 in S_1, or that it is a *tangent* to k^2.

Similarly, in the theorem on the right the point P_1 (Fig. 32) is the only point of u_1 through which there passes but one ray of K^2, namely, the ray u_1 itself. For through any other point A_1

of u_1 there passes a second ray A_1A of K^2, since A cannot coincide with P and hence A_1A differs from u_1. We call P_1 therefore a *point of contact* of the sheaf K^2 in the ray u_1. Hence :

To the common ray of two projective sheaves of rays there corresponds in each sheaf a tangent to the curve of the second order which is generated by the sheaves.	To the common point of two projective ranges of points there corresponds in each range a point of contact of the sheaf of the second order which is generated by the ranges.

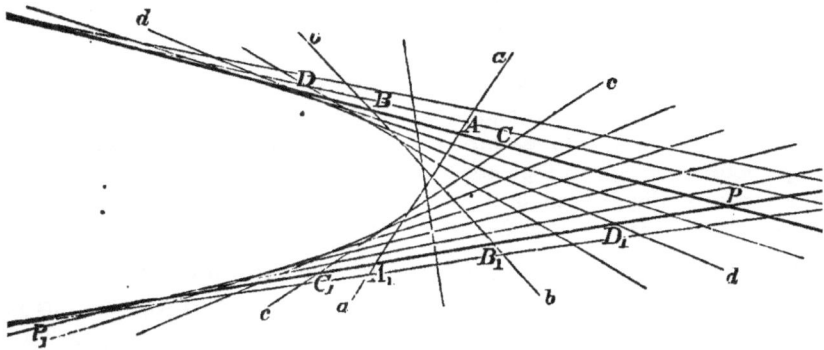

FIG. 32.

103. As you know, two one-dimensional primitive forms can be so correlated projectively to each other that to any three elements of the one correspond three elements of the other, chosen arbitrarily, and the correlation so established is unique.

If, then, we wish to construct a curve of the second order by means of projective sheaves of rays, we may not only choose at random the centres S and S_1 of the generating sheaves (Fig. 31), but also three of the points of the curve, namely, the points of intersection of three pairs of corresponding rays of the sheaves. In case one of these three points of intersection coincides with S, the tangent to the curve at the point S may be chosen arbitrarily, and the same thing is true for S_1.

In order to construct a sheaf of rays of the second order by means of projective ranges of points, we may not only choose at will the bases u and u_1 of the generating ranges (Fig. 32), but also in the plane containing u and u_1, three other rays of the sheaf, namely, the lines joining any three pairs of corresponding points of the ranges. In case one of these joining lines coincides with u, then the point of contact of the sheaf in the ray u may be chosen at will, and the same thing is true for u_1.

Consequently the problems—

To construct a curve of the second order of which five points, or four points and the tangent at one of them, or three points and the tangents at two of them, are given,	To construct a sheaf of rays of the second order of which five rays, or four rays and the point of contact in one of them, or three rays and the points of contact in two of them, are given,

admit of solution, and evidently reduce to the following problems :

Two projective sheaves of rays S and S_1 are given by three pairs of corresponding rays a and a_1, b and b_1, c and c_1 ; to construct any required number of points of the curve of the second order k^2 which these sheaves generate.	Two projective ranges of points u and u_1 are given by three pairs of corresponding points A and A_1, B and B_1, C and C_1 ; to construct any required number of rays of the sheaf of the second order K^2 which these ranges generate.

104. The solution of these problems depends upon finding the element of one primitive form which corresponds to an arbitrarily chosen element of the other when three pairs of corresponding elements are given, since at the same time a new element of the resulting form of the second order is determined. I might refer you, then, to the construction given in the last lecture (Art. 91).

I shall, however, solve the problem again and in a more symmetrical manner, particularly since many important theorems are connected with the solution. This solution consists in finding a third primitive form which is perspective to each of the given forms. Thus : *

Through the point of intersection aa_1 of any two corresponding rays of the projective sheaves S and S_1 (Fig. 33) draw two straight lines u and u_1, of which the first u cuts the sheaf $S(abc)$ in a range of points $u(ABC)$, and the second u_1 cuts the sheaf $S_1(a_1b_1c_1)$ in a	In the line AA_1 joining any two corresponding points A and A_1 of the projective ranges of points u and u_1 (Fig. 34) choose centres S and S_1 of two sheaves of rays of which the first $S(abc)$ projects the range of points $u(ABC)$, and the second $S_1(a_1b_1c_1)$ projects the range

---

* I would again urge that the beginner draw the diagram for himself according to the statement of the text, since by this means the comprehension of the solution is greatly simplified.

range of points $u_1(A_1B_1C_1)$. As sections of projective sheaves the ranges of points u and u_1 are likewise projective to each other. But they are, moreover, perspective, since in their point of intersection two homologous elements A and A_1 coincide (Art. 86). They are thus sections of a single sheaf of rays S_2 in whose centre the rays BB_1 and CC_1 intersect.

In order now to find for any ray d of the sheaf S the corresponding ray d_1 of S_1, we project the point of intersection du or D from the centre S_2 upon the straight line u_1 at D_1; then D_1S_1 is the required ray d_1, and the point dd_1 or P lies upon the curve of the second order k^2.

$u_1(A_1B_1C_1)$. As projectors of projective ranges of points these sheaves of rays S and S_1 are likewise projective to each other. But they are, moreover, perspective, since in the line SS_1 joining their centres two homologous rays a and a_1 coincide (Art. 86). They are thus projectors of a single range of points u_2, upon which the points of intersection bb_1 and cc_1 lie.

In order now to find for any point D of u the corresponding point D_1 of u_1, we determine the intersection of the straight line DS or d with u_2, and project this point of intersection from S_1 by the ray d_1 upon the straight line u_1. The intersection of d_1 with u_1 is the required point D_1, and the ray DD_1 belongs to the sheaf of the second order K^2.*

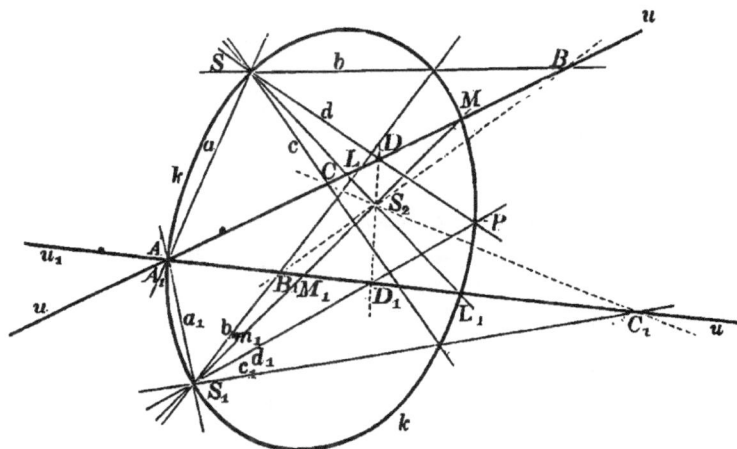

FIG. 33.

105. The following problems are solved by the same constructions :

* In Fig. 34 the sheaf of the second order K^2 is indicated by the curve which it envelops.

Upon any ray of S (or S_1) to find the second point of intersection with the curve of the second order, *i.e.* the point different from S (or S_1).

Through any point of u (or u_1) to draw the second ray of the sheaf of the second order, *i.e.* the ray different from u (or u_1).

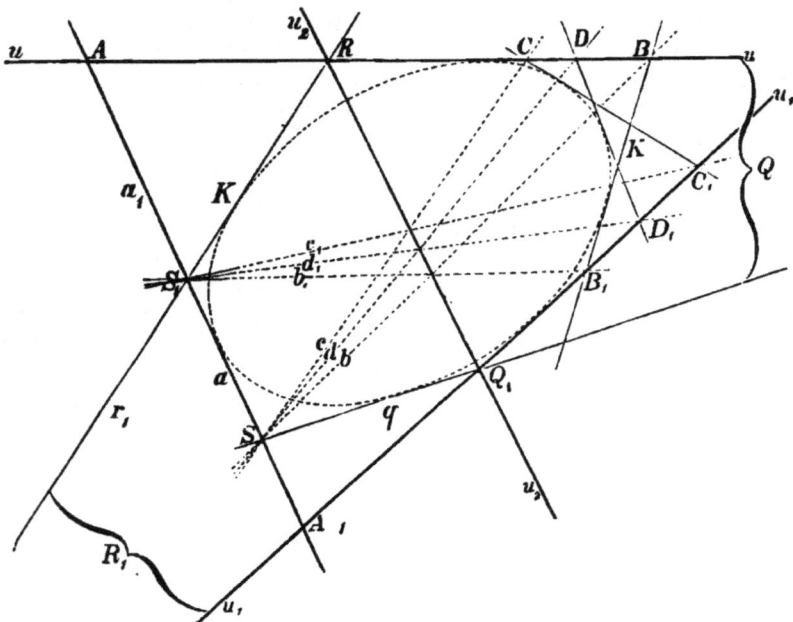

FIG. 34.

If by the same method we construct the two rays (or points) which correspond to the common ray of the sheaves (or to the point of intersection of the ranges of points) we obtain a solution for the problems :

To construct the tangents to a curve of the second order, at the centres of the generating sheaves of rays.

To find the points of contact in a sheaf of rays of the second order, upon the two projective ranges of points which generate the sheaf.

106. Another very important result is derived from the above construction.

If in the construction on the left (Fig. 33) we draw the ray S_1S_2. or m_1, which is intersected by u_1 and u in M_1 and M respectively, there corresponds to this in S the ray SM; for if D_1 is brought into the position M_1, then will both D and P coincide with M. Thus M is that point in which the curve k^2 is intersected a second time by u. Similarly the point L_1, in which u_1 cuts the curve k^2 a second time, lies upon the straight line SS_2. I would remind you at this point that the arbitrarily chosen straight lines u and u_1 are subject to no other condition than that they intersect in a point of the curve k^2.

On the other hand, if in the construction on the right (Fig. 34), we join the point of intersection u_1u_2 or Q_1 with S, the point Q of u which corresponds to Q_1 of u_1 will lie upon this straight line; or SQ_1 is a ray of the sheaf of the second order K^2; and similarly, the ray S_1R which projects the point of intersection of u and u_2 from S_1 is a ray of K^2. It will be remembered that S and S_1 were chosen arbitrarily upon a ray a of the sheaf K^2. These phases of the general constructions give us the solution of the problems :

Upon any straight line u which cuts a curve of the second order k^2 in a given point A, to find the second point of intersection with the curve.

Through any point S which lies upon a given ray a of a sheaf of the second order K^2, to draw the second ray of this sheaf.

107. Moreover, the two exceedingly fruitful theorems of Pascal and Brianchon, of which I made mention in the Introduction, arise spontaneously out of these constructions.

If, on the left, from the five points S, S_1, A, M, and L_1 of the curve k^2, we determine any sixth point P (Fig. 33), then in the construction necessary to determine P it is seen that D the point of intersection of SP and u lies in a straight line through S_2, with D_1 the point of intersection of S_1P and u_1. But D, D_1, and S_2 are those points in which the three pairs of opposite sides of the hexagon SPS_1MAL_1 intersect.

If on the other hand, in the problem on the right (Fig. 34), from the five rays u, u_1, SS_1, SQ_1, and S_1R of the sheaf of the second order K^2, we construct any sixth ray DD_1, then the straight lines SD and S_1D_1 must intersect upon the straight line Q_1R or u_2. But these three straight lines passing through one and the same point are the principal diagonals, *i.e.* the lines joining opposite vertices, of the hexagon $SS_1RDD_1Q_1$.

We accordingly have the following theorems :

PASCAL'S THEOREM.*

In any simple hexagon which is inscribed in a curve of the second order, the three pairs of opposite sides intersect in three points of one straight line.

BRIANCHON'S THEOREM.

In any simple hexagon which is formed of six rays of a sheaf of the second order, the three principal diagonals intersect in one point.

108. To be made perfectly rigorous the demonstration of these theorems must, I admit, be freed from certain restrictions. The two hexagons which here come into question contain elements which are not chosen freely ; for on the left, for example, the vertices A, M, L_1, and P were chosen at random upon the curve k^2, but the same is not true of S and S_1. It might be thought that the centres S and S_1 of the projective sheaves of rays by which the curve is generated are distinguished from other points of the curve by peculiar properties ; for example, that Pascal's theorem holds true only for those inscribed hexagons of which S and S_1 are two vertices. We shall now remove this possibility by demonstrating that any two points of the curve whatsoever might be taken as the centres of projective sheaves of rays which generate the curve, and hence that S and S_1 may be replaced in the Pascal hexagon by any two points of the curve chosen arbitrarily. The same thing is true of the straight lines u and u_1 which appear as sides in the Brianchon hexagon.

109. If in the Pascal hexagon SPS_1MAL_1 (Fig. 33) we imagine all the vertices except A to remain fixed while A moves along the curve, then L_1A or u_1 will rotate about L_1 and MA or u about M, while the points D_1 and D move upon the fixed straight lines d_1 and d in such a way that the straight line DD_1 always passes through the fixed point S_2. The Pascal theorem therefore holds true for any one of the hexagons so constituted.

Now the points D_1 and D describe two perspective ranges d_1 and d, each of which is a section of the sheaf of rays S_2. At the same time u_1 and u describe about L_1 and M, respectively, projective sheaves of rays, these being projectors of the perspective ranges of points d_1 and d ; we may thus consider the curve k^2 to

---

* Pascal discovered this fundamental property of six points of a conic section in 1639 when only 16 years of age. Brianchon published his equally fundamental theorem in 1806 in the *Journal de l'Ecole polytechnique*, Vol. XIII.

be generated by the projective sheaves of rays L_1 and M, whose centres have been chosen arbitrarily upon the curve.

Similarly, let us imagine the side SS_1 of the Brianchon hexagon $SS_1RDD_1Q_1$ (Fig. 34) to so move as to remain a ray of the sheaf K^2 while the other sides are unaltered; then S_1 describes a range of points S_1R or r_1, and S describes a range of points SQ_1 or q projective to r_1. For the point of intersection of the principal diagonals S_1D_1 and SD moves upon the fixed straight line Q_1R and describes in this line a range of points u_2 to which q and r_1 are perspective. We might therefore consider the sheaf of the second order to be generated by the projective ranges of points q and r_1.

Hence the Pascal and Brianchon theorems are perfectly general, the proof of which has brought us to the following fundamental theorems upon curves and sheaves of rays of the second order :

A curve of the second order is projected from any two of its points by projective sheaves of rays, those pairs of rays being homologous which pass through the same point of the curve.

A sheaf of rays of the second order is cut by any two of its rays in projective ranges of points, those pairs of points being homologous which lie upon the same ray of the sheaf.

110. We shall make use of these theorems later in correlating forms of the second order to each other and to the one-dimensional primitive forms, in a way similar to that in which we have established projective relations among the latter forms. With this in view and upon the basis of these theorems we propose the following definitions :

Four points of a curve of the second order are called 'harmonic points' if they are projected from any, and consequently from every, fifth point of the curve by four harmonic rays.

Four rays of a sheaf of the second order are called 'harmonic rays' if they are intersected by any, and consequently by every, fifth ray of the sheaf in four harmonic points.

Thus by three points of a curve, or three rays of a sheaf, of the second order, the fourth harmonic is determined unambiguously, and can easily be constructed as soon as it is specified from which of the three given points or rays it is to be separated.

111. Bearing in mind the statement of Article 102 concerning tangents of the curve of the second order and points of contact

of the sheaf of rays of the second order, and the results of Article 109, we conclude that—

through any point of a curve of the second order there passes one tangent to the curve.	upon any ray of a sheaf of the second order there lies one point of contact of the sheaf.

Every curve of the second order, then, is enveloped by a system of tangents, and every sheaf of rays of the second order envelops a series of points of contact. It will be one of the problems of my next lecture to show you that this system of tangents to a curve of the second order is nothing else than a sheaf of rays of the second order, and that the series of points of contact in a sheaf of rays of the second order constitutes a curve of the second order.

112. Other very important properties of curves and sheaves of rays of the second order are stated in the following theorems, of which we shall hereafter make frequent use :

Two curves of the second order coincide if they have in common either five points, or four points and the tangent at one of them, or three points and the tangents at two of them.	*Two sheaves of rays of the second order coincide if they have in common either five rays, or four rays and the point of contact in one of them, or three rays and the points of contact in two of them.*
For, if we project both curves from one of their common points S by a single sheaf of rays, to this is projective each of the two sheaves projecting the curves separately from another common point S_1. But the latter sheaves are identical in any of the three cases mentioned, since in each case they have three self-corresponding rays, namely : (1) the rays joining S_1 to the three common points of the curves, different from S and S_1; (2) the rays joining S_1 to the two remaining common points and to S if the curves have a common tangent at S ; and (3) the ray joining S_1 to the third common point, the ray S_1S and the tangent at S_1 if the curves have common tangents	For, to the range of points in which one of the common rays u intersects both sheaves of rays, is projective each of the ranges of points in which the two sheaves of rays are separately intersected by a second common ray u_1. But the latter ranges of points are identical in any of the three cases mentioned, since in each case they have three self-corresponding points, namely : (1) the points of intersection of u_1 with the three common rays of the sheaves, different from u and u_1; (2) the points of intersection of u_1 with the two remaining common rays and with u if the sheaves have a common point of contact in u ; and (3) the points of intersection of u_1 with the

at S and S_1. Every ray of S therefore intersects both curves a second
time in the same point, and hence
the two curves are identical.

third common ray, and with u and
the point of contact in u_1 if the
sheaves have common points of
contact in u and u_1. The rays of
the two sheaves, therefore, which
pass through the same point of u
coincide, and hence the two sheaves
are identical.

EXAMPLES.

1. Given five points of a curve of the second order ; with the aid of
Pascal's theorem determine upon any straight line passed through one
of them, its second point of intersection with the curve, and so construct
any required number of points of the curve.

2. Given five rays of a sheaf of the second order ; with the aid of
Brianchon's theorem determine the second ray of the sheaf passing
through any point of one of the given rays, and so construct any
required number of rays of the sheaf.

3. Prove that the circle is a curve of the second order, and that its
system of tangents constitutes a sheaf of rays of the second order.

What angle does that segment of a movable tangent to a circle,
which lies between two fixed tangents, subtend at the centre ? Is the
angle constant ?

4. A variable triangle ASA_1 so moves in its plane that the extremities of the base A and A_1 continually lie upon straight lines u and u_1
respectively, while the sides SA and SA_1 rotate about the fixed point S,
the angle S remaining of constant magnitude. Show that the base
AA_1 generates a sheaf of rays of the second order, to which u and u_1
belong. [It will be seen later that the point S is a focus of the curve
enveloped by the sheaf of rays.]

5. If two pairs of straight lines a, b and a_1, b_1 lying in the same
plane rotate, the first pair about S, the second about S_1, so that the
angles (ab) and (a_1b_1) remain of fixed magnitude, and one point of
intersection aa_1 of a pair of sides traverses a straight line, then each
of the three remaining points of intersection ab_1, a_1b, and bb_1 describes
a curve of the second order passing through S and S_1. [Newton's
organic method of describing a conic section.]

6. If two concentric sheaves of rays, whose planes intersect at an
oblique angle, are so correlated to each other that every ray is perpendicular to its homologous ray, then will they generate a sheaf of
planes of the second order ; in other words, if a right angle be
rotated about its vertex, so that one side moves in a certain fixed

plane and the other side in a different plane, the plane of the angle will envelop a cone of the second order, to which the two fixed planes are tangent.

7. If from any point perpendiculars be let fall upon the tangent planes of a cone of the second order, these will lie upon another cone of the second order.

The tangent planes are generated by two projective sheaves of rays, and from these can be immediately derived two projective sheaves of planes of the first order which generate the second cone.

8. The geometrical locus of a point S, from which a plane quadrangle $KLMN$ is projected by a harmonic sheaf of rays $S(KLMN)$, is a curve of the second order circumscribing the quadrangle (Art. 110).

If we draw the fourth harmonic ray to NK, NL, and NM, this will be tangent to the required curve at N, and hence the curve can easily be constructed.

Ascertain whether or not there is more than one curve of the second order satisfying the given conditions.

9. State the reciprocal of Example 8.

Remark: With Von Staudt we shall call a group of four elements of a one-dimensional primitive form, taken in a definite order, a 'throw' [*Wurf*]. Two throws, $ABCD$ and $abcd$, are said to be projective when the two primitive forms in which they lie can be so related projectively to each other that the elements A, B, C, D of the one correspond to a, b, c, d of the other. Example 8 may then be generalized as follows:

Suppose $abcd$ is a given throw consisting, say, of four rays of a sheaf of the first order; then all points S, from which a quadrangle $KLMN$ is projected by a sheaf $S(KLMN)$ projective to $abcd$, lie upon a curve of the second order circumscribing the quadrangle.

How would the tangent at N be constructed in this general case?

The reciprocal theorem may be generalized in a similar way.

10. If two triangles ABC and $D_1E_1F_1$ are inscribed in a curve of the second order k^2, they are also circumscribed to another curve of the second order; and conversely.

The throws $A(BCE_1F_1)$ and $D_1(BCE_1F_1)$, are projective since they consist of corresponding elements in the projective sheaves of rays A and D_1 which generate the curve. If, now, $B_1C_1E_1F_1$ is the section of the sheaf $A(BCE_1F_1)$ made by the line E_1F_1, and $BCEF$ that of the sheaf $D_1(BCE_1F_1)$ made by the line BC, then are also $B_1C_1E_1F_1$ and $BCEF$ projective throws, and the six sides of the triangles, BC, BB_1, CC_1, E_1F_1, EE_1, and FF_1, are rays of a sheaf of the second order. The converse may be proved analogously.

F

LECTURE VII.

DEDUCTIONS FROM THE THEOREMS OF PASCAL AND BRIANCHON.

113. The important properties which have just been proved concerning the hexagon in the curve and in the sheaf of rays of the second order bring us to other theorems no less important concerning pentagons, quadrilaterals, and triangles, similarly described. I must preface the deduction of these theorems with a remark upon the tangents to the curve, and the points of contact in the sheaf of rays.

114. Any ray p_1 which lies in the plane of a curve of the second order and which has only one point S_1 in common with it (Fig. 31), has been called a 'tangent' to the curve at the point S_1, and we have found that through each point of the curve a single tangent may be drawn. Any other ray a_1 of the plane passing through S_1 cuts the curve in a second point A. If now we rotate the ray a_1 about S_1, its intersection A moves along the curve and approaches indefinitely near to the point S_1, while a_1 approaches indefinitely near to the position of the tangent p_1. The *tangent* thus presents itself as *the limiting position of a straight line which joins two points of the curve indefinitely near (consecutive) to each other*, and this definition clearly applies to tangents not only of curves of the second order, but of any curves whatsoever. Similarly, we have named any point P_1 (Fig. 32), through which passes only one ray u_1 of a sheaf K^2 of the second order, a 'point of contact' of the sheaf in the ray u_1, and have found that upon each ray of the sheaf K^2 there lies one, and only one, point of contact. Through any other point A_1 of u_1 there passes a second ray a of the sheaf. If we move A_1 along u_1, a traverses the sheaf

K^2, and approaches indefinitely near to the position of the ray u_1 as A_1 approaches indefinitely near to the point P_1. In this way the *point of contact* appears as *the limiting position of the point of intersection of two rays of the sheaf indefinitely near to each other.*

115. If, then, in a hexagon which is inscribed in a curve of the second order, two adjacent vertices approach indefinitely near to each other, the side joining them assumes the position of a tangent to the curve; and if in a hexagon whose sides are rays of a sheaf of the second order, two adjacent sides approach indefinitely near to each other, in the place of the vertex in which they intersect we have a point of contact of the sheaf. The hexagon will become a pentagon, a quadrangle, or a triangle, according

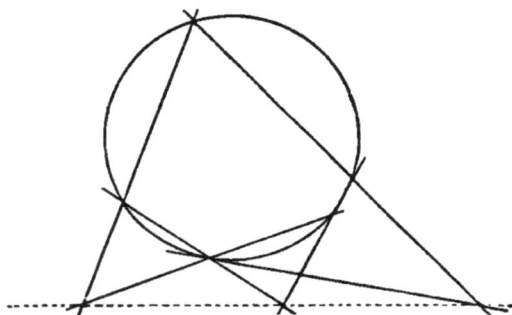

FIG. 35.

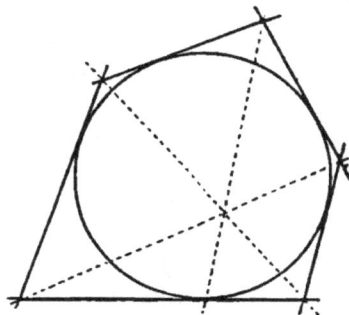

FIG. 36.

as one, two, or three pairs of adjacent elements coincide. As applied to the pentagon, the theorems of Pascal and Brianchon read as follows:

In any pentagon inscribed in a curve of the second order, the points of intersection of two pairs of non-adjacent sides lie in a straight line with that point in which the fifth side is cut by the tangent at the opposite vertex.

In any pentagon formed from rays of a sheaf of the second order, the two straight lines which join two pairs of non-consecutive vertices intersect that straight line which joins the fifth vertex to the point of contact of the opposite side in one and the same point.

This double theorem affords the solution of the following two problems:

Given any five points of a curve of the second order, to draw the tangents at these points with the use of the ruler only.

Given any five rays of a sheaf of the second order, to find their points of contact with the use of the ruler only.

116. For the quadrangle and quadrilateral we obtain the following theorems (Fig. 37) :

In any quadrangle inscribed in a curve of the second order, the points of intersection of pairs of opposite sides lie in a straight line with the points of intersection of the tangents at opposite vertices.

In any quadrilateral formed of rays of a sheaf of the second order, the diagonals and the straight lines joining the points of contact in opposite sides intersect in one point.

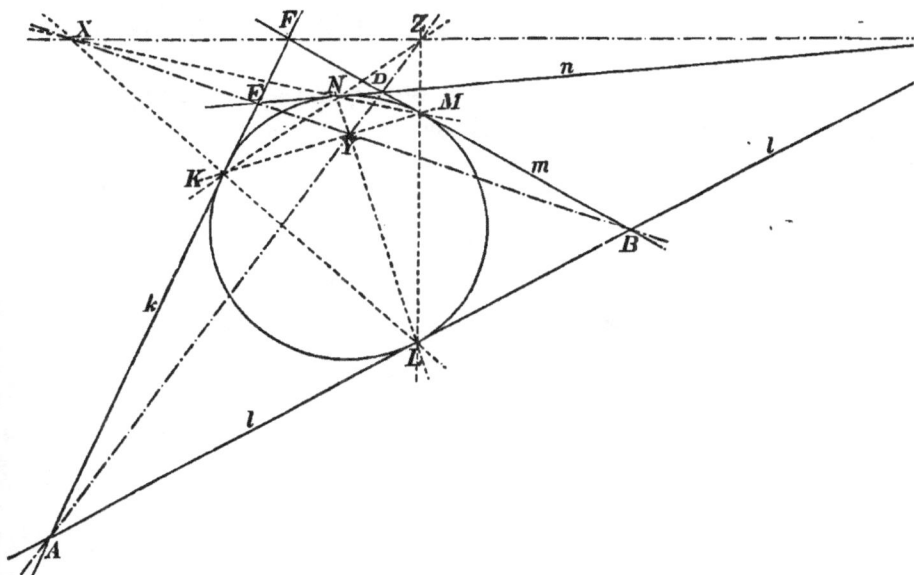

FIG. 37.

117. And finally for the triangle we have the following :

The three points in which the sides of any triangle inscribed in a curve of the second order are intersected by the tangents at the opposite vertices lie upon one straight line.

In any triangle whose sides are rays of a sheaf of the second order, the three straight lines which join the vertices to the points of contact on the opposite sides intersect in one point.

118. All these theorems, which prove very useful for the solution of a series of simple problems (in particular those of Art. 103), admit of direct deduction without reference to the hexagon. I shall by way of illustration give you the direct proof for the theorems upon the quadrangle, since this proof discloses important new properties of projective primitive forms, and since I shall make use of these in deducing other results.

In order readily to find in one of two projective sheaves S and S_1 (Fig. 33) the ray corresponding to any chosen ray of the other, we construct (Art. 10⬤) a third sheaf of rays S_2 perspective to each of the given sheaves. For this purpose we intersect S and S_1 in the ranges of points u and u_1, respectively, by two straight lines passing through A the point of intersection of any two homologous rays a and a_1 of the given sheaves ; since these two ranges are perspective, the sheaf S_2 of which they are sections is the one required.

This is equally true if u coincides with a_1 and u_1 with a (Fig. 38), so that the points ba_1 and b_1a, or B and B_1, ca_1 and c_1a, or

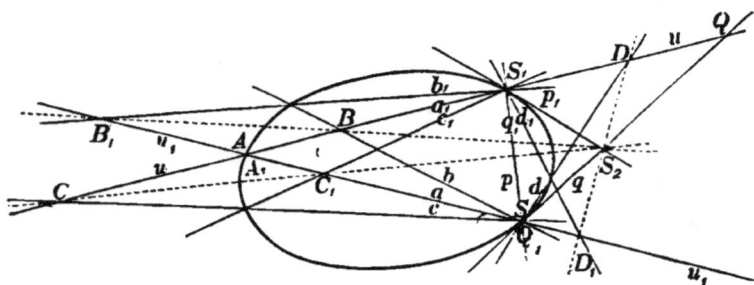

FIG. 38.

C and C_1, etc., in which two corresponding rays a and a_1 are cut by any two others, as b and b_1, c and c_1, alternately, lie in a straight line with a fixed point S_2. If through this point S_2 a straight line DD_1 is passed which cuts the rays a_1 and a in the points D and D_1, respectively, then are SD and S_1D_1 corresponding rays of the sheaves S and S_1. If now we bring D_1 into coincidence with S, SD falls upon SS_2 or q, so that to this ray of S corresponds the ray S_1S or q_1, which joins the centres of the sheaves S and S_1. The straight line SS_2 or q is thus a tangent to the curve of the second order generated by S and S_1, and similarly S_1S_2 or p_1 is also a tangent. Consequently the fixed point S_2 is the intersection of the two tangents q and p_1 drawn at S and S_1, respectively, *i.e.* the intersection of the two rays which correspond (in the two sheaves) to the common ray SS_1. We obtain therefore the same point S_2 whether we let u_1 and u coincide respectively with a and a_1 or with some other pair of homologous rays (as b and b_1 or c and c_1). That is to say : The straight line which joins the points in which any two pairs whatsoever of homologous rays intersect alternately (*e.g.*, ab_1 and a_1b, ac_1 and a_1c, etc.), passes through the point S_2.

On the other hand, in order easil● to find the point in one of two projective ranges u and u_1 (Fig. 34) which corresponds to any point of the other, we determine (Art. 104) a third range u_2 which is perspective to each of the given ranges in the following way : Project u and u_1 by two sheaves of rays S and S_1 whose centres are chosen upon the straight line a joining any two homologous points A and A_1 of the given ranges ; since these sheaves are perspective, the range u_2 of which they are both projectors is the one sought.

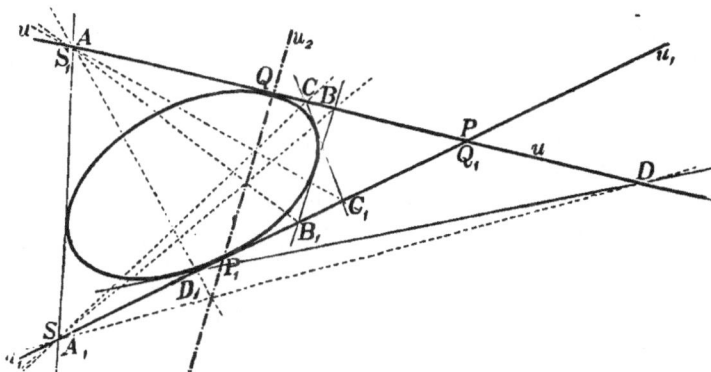

FIG. 39.

If now we let S coincide with A_1 and S_1 with A (Fig. 39), u_2 will pass through those two points of u and u_1 which correspond to their point of intersection. For, two arbitrary points D and D_1 which correspond to each other are projected from A_1 and A respectively by two rays A_1D and AD_1 which intersect upon the straight line u_2 ; but obviously this intersection, and consequently also D_1, coincides with the point u_1u_2 or P_1 if D is brought into coincidence with uu_1 or P, so that u_2 passes through one (and similarly through the other) of the two points P_1 and Q which correspond to the intersection point P, Q_1, of u and u_1. In other words, u_2 joins the points of contact of the sheaf generated by u and u_1, lying on these two lines. We therefore always obtain the same straight line u_2 whether we let the centres S_1 and S coincide with A and A_1 or with some other pair of homologous points, as B and B_1 or C and C_1. The point of intersection of any two straight lines joining pairs of homologous points alternately (*e.g.*, the lines BC_1 and B_1C) lies upon the straight line u_2.

119. The results just obtained may be arranged in the following double theorem so as to show their relation to each other :

The two points ab_1 and a_1b, in which any two pairs a, a_1 and b, b_1 of homologous rays in projective sheaves S and S_1 intersect alternately, lie in a straight line with the point S_2, in which the rays corresponding to the common ray SS_1 intersect.	The two straight lines AB_1 and A_1B, which join any two pairs of homologous points A, A_1 and B, B_1 of projective ranges u and u_1 alternately, intersect upon the straight line u_2 which joins those points of the ranges corresponding to the intersection point of u and u_1.

120. You will immediately recognize in these theorems those already stated concerning the quadrangle in the curve and in the sheaf of the second order from the following remarks :

In the curve of the second order, which is generated by the sheaves S and S_1, the points S, aa_1, S_1, and bb_1 determine an inscribed quadrangle in which a and b_1, also a_1 and b, are opposite sides, while the two tangents which touch the curve in the opposite vertices S and S_1 intersect in S_2.	In the sheaf of rays of the second order, which is generated by the ranges of points u and u_1, the rays u, AA_1, u_1, and BB_1 form a quadrangle in which A and B_1, also A_1 and B, are opposite vertices, while the two points of contact in the opposite sides u and u_1 lie upon u_2.

It is apparent that the theorems of the last Article afford a very convenient method of determining that element in one of two projective primitive forms of one dimension which corresponds to any given element of the other. For example, if in two projective ranges of points u and u_1 (Fig. 39) three pairs of corresponding points are given, the straight line u_2 is obtained immediately from these, and this in turn very simply determines the point of u_1 corresponding to any given point of u.

121. The theorems upon the quadrangle in the curve and in the sheaf of rays of the second order, proved here for a second time, may be stated in the following general form :

If four points K, L, M, N, *of a curve of the second order determine*	*If four rays* k, l, m, n, *of a sheaf of the second order determine a*

a complete quadrangle and their complete quadrilateral and their tangents k, l, m, n, a complete quad- points of contact a complete quad- rilateral, the three pairs of opposite rangle, the three pairs of opposite vertices of the quadrilateral lie sides of the quadrangle pass through upon the straight lines joining the the intersection points X, Y, Z, of points X, Y, Z, in which pairs of the straight lines (diagonals) join- opposite sides of the quadrangle ing pairs of opposite vertices of the intersect (Fig. 37). quadrilateral (Fig. 37).

For, the theorems as previously stated are true for each of the three simple quadrangles comprised within the complete quadrangle *KLMN* on the left, and for each of the three simple quadrilaterals comprised within the complete quadrilateral *klmn* on the right.

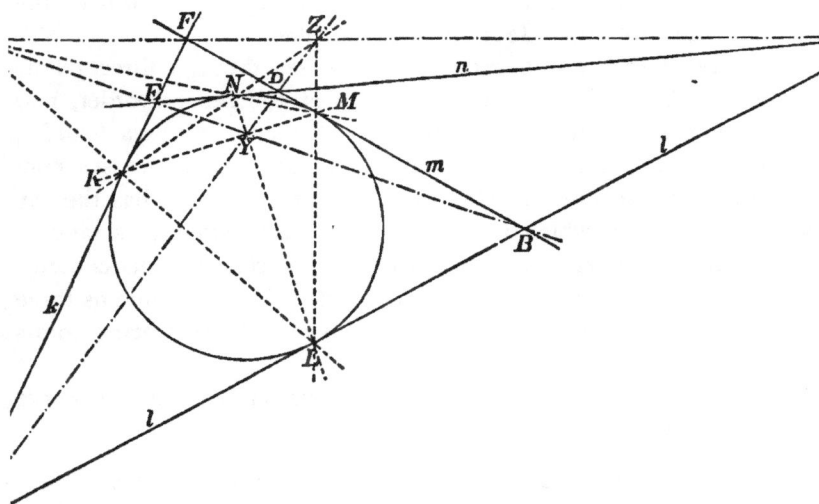

FIG. 37.

But in this form the theorem on the left states exactly the same thing as that on the right ; for both state that the triangle whose sides join the pairs of opposite vertices of the quadrilateral is identical with the triangle *XYZ*, in whose vertices the pairs of opposite sides of the quadrangle intersect.

And conversely, if a quadrangle *KLMN* have such position relative to a quadrilateral *klmn* circumscribing it, then a curve of the second order can be constructed which touches the straight lines *k, l, m, n,* in the points K, L, M, N; or, a sheaf of rays of the second order can be constructed which has the points K, L, M, N, as points of contact in the rays k, l, m, n. For in consequence of our theorem a curve of the second order which passes through the points K, L, M, N, and touches the straight line k in K has also the straight lines l, m, n, as tangents ; and the sheaf of the second order, which contains the rays k, l, m, n, and has K as point of contact in k, has also L, M, N, as points of contact. Therefore four tangents of a curve of the second order and their four points of contact may always be looked upon as four rays of a sheaf of the second order and their four points of contact.

A curve of the second order, which passes through three points of contact K, L, M, of a sheaf of rays of the second order, and in two of these points K and L has as tangents the rays k and l, respectively, to which these points of contact belong, passes consequently through every fourth point of contact N, and has the ray of the sheaf n, to which this point of contact belongs, as tangent in this point. Conversely, a sheaf of rays of the second order contains every tangent of a curve of the second order if only it contains three of these tangents, the points of contact in two of them being points of the curve.

We thus have the following beautiful relation between the curve and the sheaf of rays of the second order :

The tangents of a curve of the second order form a sheaf of rays of the second order.	*The points of contact of a sheaf of rays of the second order form a curve of the second order.*

122. On account of their importance we shall prove these relations again, and by a method in which a new and interesting property of the curve of the second order is brought to light.

Of the four vertices K, L, M, N (Fig. 37), of a quadrangle inscribed in a curve of the second order, let any one of them, K for example, move upon the curve, while the remaining three and their tangents do not alter their positions. Then the tangent k of the point K glides along the curve, and its points of intersection, E and A, with the tangents n and l, move along these tangents. It is easy to see that by this motion E and A describe two projective ranges of points upon n and l, respectively, and hence that the tangent k describes a sheaf of rays of the second order. For the two diagonals EB and AD of the quadrilateral k, l, m, n, intersect always in a point Y of the fixed straight line LN, and describe in consequence, about the fixed vertices B and D, two perspective sheaves of rays; the ranges of points n and l described by the points E and A are sections of these, and are therefore projective. Hence the moving tangent generates a sheaf of rays of the second order. That the points of contact in a sheaf of rays of the second order form a curve of the second order may be proved in a similar manner.

The straight line MK always passes through the point Y, and, by the motion of the point K, describes a sheaf of rays about the fixed point M, which is perspective to the sheaf B described by BE, and hence is projective to the range of points n described by E. From this it follows as K traverses the whole curve:

Of a curve of the second order if there be given any fixed point M, *and any fixed tangent* n, *and we correlate to each ray through* M, *projecting a point* K *of the curve, that point of* n *through which passes the tangent at the point* K, *then the sheaf of rays* M *is projectively related to the range of points* n.

This theorem, which Chasles places at the beginning of his treatise on conic sections, is, in the plane, self-reciprocal, since, as was just now shown, or as follows directly from the theorem itself, the tangents to a curve of the second order form a sheaf of rays of the second order.

I shall only add that—

The tangents at four harmonic points of a curve of the second order are harmonic tangents.

That is to say, they are cut by any fifth tangent in four harmonic points, since their four points of contact are projected from any fifth point of the curve by four harmonic rays.

The fact that a sheaf of the second order is cut by any two of its rays in projective ranges of points may now be stated thus:

"The sheaf of rays formed by the tangents to a curve of the "second order is cut by any two of these tangents in projective "ranges of points."

123. The reciprocal relation existing between the Brianchon theorem and the Pascal theorem may be seen directly when the former is stated as follows:

"The three principal diagonals of any hexagon circumscribed "to a curve of the second order (*i.e.* whose sides are tangent to "the curve) intersect in one point."

So also the theorems upon the pentagon, the quandrangle, and the triangle in a sheaf of rays of the second order may be conveniently stated as referring to figures circumscribing a curve of the second order.

124. I shall not do more at this point than to reproduce in this new form a single one of the previously proved theorems relating to forms of the second order lying in a bundle of rays, and its reciprocal. It has already been shown (Art. 100) that any curve or any sheaf of rays of the second order is projected from a point not lying in the same plane by a cone or by a sheaf of planes of the second order. Every tangent to the curve is projected by a plane which has but one ray in common with the cone, and is therefore called a 'tangent plane.' Similarly every point of contact of the sheaf of rays of the second order is projected by a so-called 'contact ray' of the sheaf of planes, through which passes only one plane of the sheaf. Since, then, every cone and every sheaf of planes of the second order is cut by a plane not containing its centre, in a curve or in a sheaf of rays of the second order, it follows that—

The tangent planes of a cone of the second order form a sheaf of planes of the second order.

The rays of a cone of the second order are projected from any two of them by projective sheaves of planes (comp. Art. 109).

The contact rays of a sheaf of planes of the second order form a cone of the second order.

The tangent planes of a cone of the second order are cut by any two of them in projective sheaves of rays.

125. As in Article 110 we may now set up the following definitions:

Four rays of a cone of the second order are called 'harmonic rays' if they are projected from any, and hence from every, fifth ray of the cone by harmonic planes.

Four tangent planes of a cone of the second order are called 'harmonic tangent planes' if they are cut by any, and hence by every, plane tangent to the cone in four harmonic rays.

The Pascal and Brianchon theorems may be stated thus for the cone of the second order:

In any hexagonal pyramid inscribed in a cone of the second order, the three pairs of opposite faces intersect in straight lines which lie in one plane.

In any hexagonal pyramid circumscribed to a cone of the second order, the three principal diagonal planes intersect in one straight line.

It would be a useful exercise to transfer for yourselves the remaining theorems which we have proved for plane figures, to the corresponding forms in the bundle.

126. At this point I shall return to the curve of the second order to draw some inferences concerning the forms which it takes in the plane, from the fact that it has not more than two points in common with any straight line. In consequence of this property such a curve has in common with the infinitely distant line of its plane either no point, or one point at which the line is tangent to the curve, or, finally, two points in which it is intersected by this line.

In the first case all points of the curve are in the finite region of the plane, and all tangents are actual rays of the plane; such a curve is called an 'ellipse' (see Figs. 33 to 39).

In the second case the curve extends indefinitely with two branches towards the point in which it is touched by the infinitely distant straight line. This form is called a 'parabola' (see Fig. 32).

In the third case the curve consists of two curved lines, each of which extends with two branches to infinitely distant points at which the two curved lines are connected; in this case the curve is called an 'hyperbola' (Fig. 31).

Since the infinitely distant straight line of the plane cuts the hyperbola, all tangents to this curve, in particular those at the two infinitely distant points, are actual rays of the plane. The tangents which touch the hyperbola in infinitely distant points are called the 'asymptotes' of the curve.

127. These three varieties of the curve of the second order may be cut from any cone of the second order whose vertex does not lie infinitely distant. A plane σ passed through the vertex has, in common with the cone, either this point alone or it touches the cone along one ray s, or, finally, it cuts the cone in two rays p and q. Any plane σ_1 parallel to σ, in the first of these cases, cuts all rays of the cone in actual points, and the cone itself in an ellipse. In the second case, the curve of intersection is a parabola since the parallel ray s is intersected in an infinitely distant point of σ_1; the line of intersection of σ_1 with the tangent plane σ is the infinitely distant tangent to the parabola. Finally, in the third case the curve of intersection is a hyperbola, since the two rays p and q are intersected by σ_1

in their infinitely distant points. The planes which are tangent to the cone along p and q are cut by σ_1 in the asymptotes to the hyperbola; the hyperbola consists of two curved lines, since both nappes of the cone are cut by σ_1. We may consider a hyperbola, like any other variety of the curve of the second order, to be a continuous closed curve, since any cone by which such a curve is projected is a continuous closed surface.

128. Two projective sheaves of rays which lie in a plane and are not perspective generate an ellipse, a parabola, or a hyperbola, according as they have no pair, one pair, or two pairs of corresponding rays parallel. If, now, one of these sheaves be given a translation in the plane so that it becomes concentric with the other without changing the direction of its rays, these concentric sheaves will have in the first case no self-corresponding rays, in the second case one, and in the third case two such rays. The third case always happens if the two sheaves are oppositely projective. Thus we may know immediately in many instances whether the curve determined by a sufficient number of given conditions, for example, by five of its points, is an ellipse, a parabola, or a hyperbola.

Investigations similar to those of the preceding paragraph become more difficult if the curve is determined by its tangents or by two

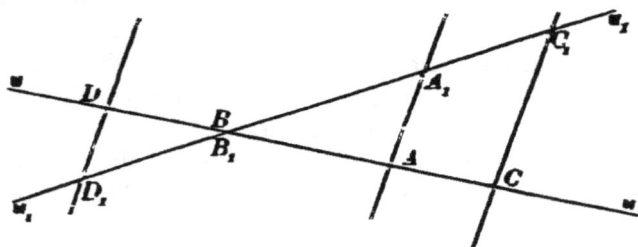

FIG. 40.

projective ranges of points which generate the enveloping sheaf of rays. However, we know immediately that two projective ranges of points generate the system of tangents to a parabola only when their infinitely distant points correspond to each other, for it is only under these conditions that the infinitely distant line of the plane becomes a tangent to the curve. We call two projective ranges of points whose infinitely distant points correspond to each other, 'similarly' projective. If two such ranges were brought into perspective position by causing any two actual points which correspond to coincide, they would appear as sections of a sheaf of parallel rays, since their centre of projection must lie upon the infinitely distant line, and therefore be infinitely distant.

Incidentally we have this metric relation for the tangents of a parabola :

" Any two tangents u and u_1 of a parabola are cut proportionally "by the remaining tangents AA_1, BB_1, DD_1, etc." (Fig. 32). It is from this property moreover that we derive the expression ' similarly projective.'

129. At this point the following theorems may be introduced :

" If the vertices of a triangle so move upon three straight lines "given in a plane that two sides of the triangle do not alter their "directions, the third side either does not alter its direction or else "it describes a sheaf of rays of the second order which envelops "a parabola."

For, two of the ranges of points determined upon the three given lines by the sheaves of parallel rays described by the first two sides of the triangle are similarly projective to the third range, and hence similarly projective to each other. The third side of the triangle joins corresponding points of these two ranges.

" If a range of points u and a sheaf of rays S which lie in the "same plane are related projectively to each other, and through each "point of u be drawn a straight line parallel to the corresponding "ray of S, these will either intersect in one point or will envelop "a parabola."

That is, if we cut the sheaf of rays S with the infinitely distant line of the plane we obtain an infinitely distant range of points which is projective to u. If this is not perspective to u it will generate with u a sheaf of rays of the second order which contains the infinitely distant straight line, and consequently envelops a parabola.

130. If we project a curve of the second order from an infinitely distant point which does not lie in its plane, we obtain a cone with an infinitely distant vertex whose rays therefore are parallel. This is called a 'cylinder' of the second order. Two projective sheaves of planes with parallel axes generate therefore either a sheaf of parallel rays or a cylinder of the second order. We distinguish the cylinder as elliptic, parabolic, or hyperbolic, according as it is cut by a plane not passing through its infinitely distant vertex in an ellipse, parabola, or hyperbola, or what is the same thing, according as it contains none, one, or two infinitely distant rays.

EXAMPLES.

1. Of a curve of the second order there are given four points and the tangent at one of them, or three points and the tangents at two of them ; through one of the given points a straight line is drawn at random ; with the aid of Pascal's theorem determine its second point of intersection with the curve.

2. Of a sheaf of rays of the second order there are given four rays and the point of contact in one of them, or three rays and the points of contact in two of them ; a point is chosen at random in one of the given rays ; with the aid of Brianchon's theorem determine the second ray of the sheaf passing through this point.

3. Construct a hyperbola of which there are given the asymptotes and one point or one tangent.

4. Construct a parabola of which there are given either four tangents, or three tangents and the point of contact in one of them, or two tangents and their points of contact.

5. Construct a hyperbola, having given three points and the *directions* of the asymptotes.

6. Given a range of points u and a sheaf of rays of the first order S lying in the same plane and projectively related ; the straight lines drawn from the points of u perpendicular to the corresponding rays of S either envelop a parabola to which u is tangent or pass through one point.

7. If an angle of given magnitude so moves in its plane that its vertex describes a straight line u while one side continually passes through a fixed point S, its other side will envelop a parabola to which u is tangent.

8. The base AA_1 of a variable triangle APA_1 is of fixed length and slides along a fixed straight line u while the other two sides AP and A_1P pass through fixed points S and S_1 respectively. The vertex P describes a hyperbola which passes through S and S_1 and has u for an asymptote.

9. Show that a parabola cannot have two parallel tangents aside from the infinitely distant line which is parallel to every tangent.

10. If the vertices of a simple hexagon $AC_1BA_1CB_1$ lie alternately upon two straight lines u and u_1, say A, B, C, upon u and A_1, B_1, C_1, upon u_1, then the points of intersection A_2, B_2, C_2, of the three pairs

of opposite sides lie upon a third straight line u_2 (Art. 119). This theorem may be found in Pappus, *Math. Coll. VII.* The diagram illustrating the theorem, like that mentioned in Art. 7 (Fig. 3), is worthy of notice on account of its regularity. It consists of nine points which lie by threes upon nine straight lines, and these nine lines pass by threes through the nine points. The diagram also illustrates the reciprocal theorem. State this.

11. If *ABCD* be any complete quadrangle whose six sides *AB*, *AC*, *AD*, *BC*, *BD*, *CD*, are cut by an arbitrary straight line *a* in the points *P*, *Q*, *R*, *S*, *T*, *V*, respectively, and if *E*, *F*, *H*, *K*, *L*, *M*, are the harmonic conjugates of these points with respect to the pairs of vertices of the quadrangle, so that *AEBP*, *AFCQ*, etc., are harmonic ranges, then a curve of the second order may be passed through the six points *E*, *F*, *H*, *K*, *L*, *M*, on which will also lie the three points of intersection *X*, *Y*, *Z*, of the pairs of opposite sides of the quadrangle [*Annals of Mathematics*, VII., p. 73].

As in Example 9, p. 51, *E*, *F*, *S* ; *E*, *H*, *T*, and all similarly situated sets of points are collinear. Hence in the hexagon *EFHMLK*, the pairs of opposite sides *EF* and *ML* intersect in *S*, *FH* and *LK* in *V*, *HM* and *KE* in *Q* ; the points *S*, *V*, *Q*, are collinear and therefore by Pascal's theorem the hexagon is inscriptible in a curve of the second order.

If *X* is the point of intersection of the opposite sides *AB* and *CD*, then in the hexagon *EHFKMX* the pairs of opposite sides *EH* and *KM*, *HF* and *MX*, *FK* and *XE*, intersect in the collinear points *T*, *V*, *P*, respectively. Hence *X*, and similarly the other two diagonal points, lies on the conic determined by the remaining five vertices of the hexagon, *i.e.* on the conic through the six harmonic conjugate points mentioned in the theorem.

State and prove the reciprocal theorem.

12. If a plane *a* cut the six edges of a tetrahedron *ABCD* in the points *P*, *Q*, *R*, *S*, *T*, *V*, respectively, and the harmonic conjugates of these points, *E*, *F*, *H*, *K*, *L*, *M*, with respect to the two vertices on the same edge, be found, then from any point *O* of the plane *a* these six points are projected by a cone of the second order upon which will also lie the three rays that can be drawn from *O* to meet a pair of opposite edges of the tetrahedron.

The proof of this theorem is analogous to that of the preceding one, and is given in the article quoted above. Special cases of both theorems are also mentioned there.

LECTURE VIII.

POLE AND POLAR WITH RESPECT TO CURVES OF THE SECOND ORDER.[*]

131. The theorems upon quadrangles inscribed or circumscribed to a curve of the second order introduce us to a series of very important properties of this curve, to one of which I have already referred in the Introduction. We derive these from the following considerations:

If in the plane of a curve of the second order there is given a point U (Figs. 41 and 42) which does not lie upon the curve, and through this point are drawn two straight lines AC and BD cutting the curve, these may be considered the diagonals of a simple quadrangle $ABCD$ inscribed in the curve. The two pairs of opposite sides AB, CD and BC, AD of this quadrangle intersect in points P and Q, the tangents a and c at two opposite vertices A and C intersect in a third point R, which three points P, Q, and R lie in one straight line u. Upon this line also lies the point of intersection of the tangents b and d, which may be constructed at the vertices B and D (compare the theorem upon the inscribed quadrangle, Arts. 116 and 121).

If now we denote the points of intersection of u with AC and BD, respectively, by V and W, we know immediately that V is

---

[*] The polar theory of conic sections is usually ascribed erroneously to De la Hire. It is in reality due to Desargues, whose development of the theory was given in his *Brouillon projet d'une atteinte* ..., in 1639 (compare Desargues *Œuvres, réunies par Poudra*, Paris, 1864, T. I.). Apollonius had previously shown (*Conicorum*, Lib. III., prop. xxxvii.) that the point of intersection of two tangents to a conic is harmonically separated from the points of the chord of contact by the curve.

G

harmonically separated from U by the points A and C. For, a pair of opposite sides of the quadrangle $ABCD$ intersect in A and in C, while the diagonal BD passes through U and the diagonal PQ through V. Similarly U and W are harmonically separated by B and D.

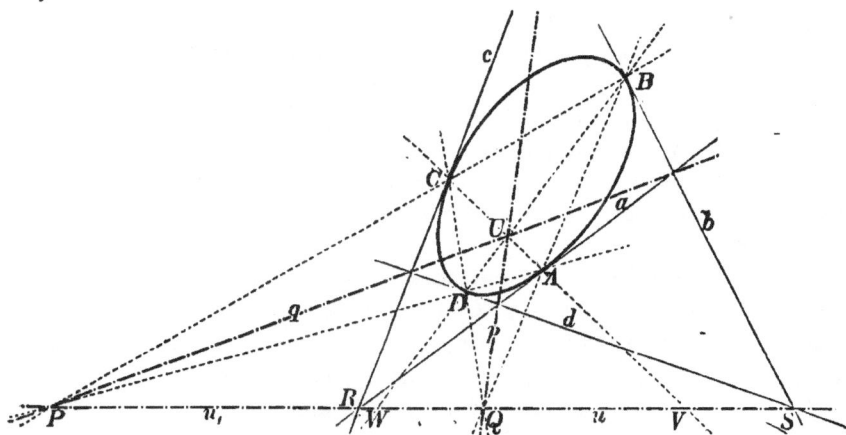

FIG. 41.

The straight line u therefore might be found by passing through U only one secant AC and determining upon it the point V which

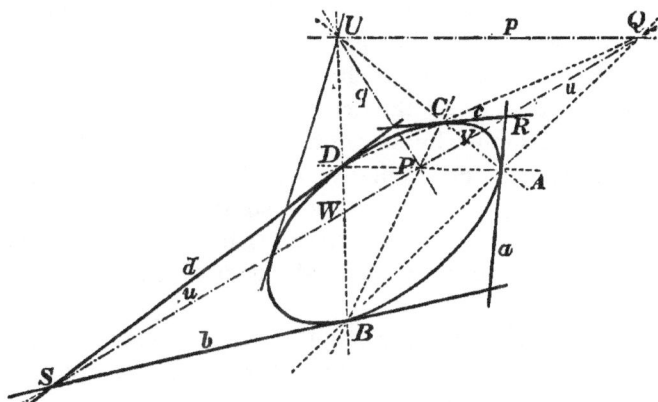

FIG. 42.

is harmonically separated from U by the points A and C of the curve, then joining V with R, the point of intersection of the tangents at A and C. In whatever position the second secant

BD may be drawn through *U,* the following points must always lie upon the straight line *u* which has already been determined by the points *V* and *R,* viz.:

(1) The points of intersection *P* and *Q* of the pairs of opposite sides of the quadrangle *ABCD*;

(2) The point of intersection *S* of the tangents at *B* and *D*;

(3) The point *W* which is harmonically separated from *U* by *B* and *D*.

We shall call *U* the *pole* of the line *u* determined as above, and conversely, the straight line *u* the *polar* of the point *U*. The polar of a given point with respect to a curve of the second order can thus be found in different ways and by linear constructions.

132. On the other hand, the pole *U* of a given straight line can be constructed by drawing from any two points *R* and *S* of the line, the two pairs of tangents *a, c* and *b, d* to the curve (Figs. 41 and 42). Through the pole *U* there pass—

(1) The diagonals of the simple quadrilateral *abcd*;

(2) The straight lines *AC* and *BD* joining the points of contact of these pairs of tangents;

(3) Two rays harmonically separated from *u*, the one by *a* and *c*, the other by *b* and *d*.

For *U* the point of intersection of *AC* and *BD* is the pole of *u*, since its polar by definition, that is, the line determined by the points of intersection of tangents at *A* and *C* and at *B* and *D*, coincides with *u*. Moreover, the straight lines *RU* and *u* are harmonically separated by the tangents *a* and *c*, since the points *U* and *V* are harmonically separated by the points of contact of these two tangents; and similarly, *SU* is harmonically separated from *u* by the tangents *b* and *d*, so that the statement (3) above is true. The correctness of statement (1) follows immediately from the theorem upon the circumscribed quadrilateral already proved (Art. 116).

133. If the polar *u* of a point *U* (Fig. 42) cuts the curve of the second order, the straight lines which join the point *U* with the two intersections are tangent to the curve at these points. For, if either of these lines had a second point in common with the curve, these two points would be harmonically separated by *U* and *u*, and the first could then not lie upon *u*. The straight line *u* is therefore the chord of contact for the point *U*, that is, the chord which joins the points of contact of the two tangents to the curve drawn from U.

134. All these results may be summarized as follows:

If through a point U which lies in the plane of a curve of the second order, but not upon the curve, any number of secants of the curve be drawn and we determine—

(1) The points of intersection of the pairs of opposite sides of any simple quadrangle inscribed in the curve which has two of these secants as diagonals;

(2) The point upon any secant which is harmonically separated from U by the points of intersection with the curve;

(3) The common point of the two tangents at the intersections with the curve of any one of these secants;

(4) The points of contact of the tangents which can be drawn to the curve from U;

Then all these points lie upon a straight line u which is called the polar of the point U with respect to the curve of the second order.

If from each one of any number of points of a straight line u which lies in the plane of a curve of the second order, but which does not touch the curve, two tangents be drawn to the curve and we determine—

(1) The diagonals of any simple circumscribing quadrilateral whose opposite sides form a pair of these tangents drawn from the same point;

(2) For any point of u, the straight line which is harmonically separated from u by the two tangents to the curve drawn from this point;

(3) The straight line which joins the points of contact of any one of these pairs of tangents;

(4) If u cuts the curve, the two tangents at the points of intersection;

Then all these straight lines pass through a point U which is called the pole of the straight line u with respect to the curve of the second order.

135. If a point A lies upon a curve of the second order, and is the tangent at this point, then a shall be called the polar of and A the pole of a. This case may be considered a limiting case of the preceding. Thus to every point in the plane of a curve of the second order there is correlated one polar with respect to the curve, and conversely, to every straight line, a pole.

136. If a point is given in the plane of a curve of the second order it is said to lie 'outside' or 'inside' the curve according as two tangents or no tangents* can be drawn from it to the curve.

---

* Only real tangents and real points of intersection are considered in these early lectures.—H.

All the points of a tangent thus lie 'outside' the curve, the point of contact alone being 'upon' the curve. Every straight line which lies in the plane of the curve contains infinitely many points which lie outside the curve, since it has one point in common with each tangent; these points evidently form a continuous series, since the tangents form a continuous series. All the points of a straight line which lie inside the curve likewise form a continuous series. If, then, the straight line joining two points which lie outside the curve cuts the curve, these two points are not separated by the points of intersection with the curve, but we can move along that segment of the line which lies outside the curve from any point *A* to any other point *B*, through the infinitely distant point if necessary, without passing over a point of the curve. Likewise two points lying inside the curve are not separated by the curve, but lie upon a segment of their joining line enclosed within the curve. On the other hand, of two points which are separated from each other by the curve, the one lies inside and the other outside the curve, for, from the preceding proof, they both can lie neither inside nor outside the curve. Hence :

"A plane is divided by a curve of the second order lying in it "into two parts. From any point of one part we can pass to any "other point of that part, but to no point of the other part, without "crossing the curve. The points of one part lie outside the curve, "and from any one of these two tangents can be drawn to the "curve ; the points of the other part lie inside the curve, and no "tangents to the curve pass through them."

Accordingly a point lying inside the curve is harmonically separated from *every* point of its polar by the curve, and *every* straight line passing through it cuts the curve in two points, while its polar is not intersected by the curve. On the other hand, a point *R* lying outside the curve is not separated from all points of its polar by the curve, but if we draw from *R* the two tangents to the curve these limit upon the polar the segment whose points are harmonically separated from *R*. The curve and all of its enclosed points are contained in one or other of the two complete angles which are formed by any two tangents.

137. We shall make frequent use hereafter of these facts, which perhaps have appeared to you to be self-evident, but which from my point of view needed demonstration. At the outset I shall prove with their help the following fundamental theorems in the polar theory.

The polars of all the points of a straight line u pass through the pole U of this line.

(1) If a point *P* of *u* lies inside the curve of the second order it is harmonically separated from *U* by the curve, and its polar *p* must therefore pass through *U*, since it contains all points which are harmonically separated from *P* by the curve.

(3) From a point *R* of *u* which lies outside the curve we can draw two tangents to the curve; the straight line which joins the point of contact of one of these tangents with *U* is the polar of *R*, since its pole must lie both upon *u* and upon this tangent.

(5) Finally, if a point of *u* lies upon the curve, the tangent at this point is its polar; but this also passes through *U*.

The poles of all the straight lines through a point U lie upon the polar u of this point.

(2) If a ray of *U* cuts the curve of the second order we determine its pole by constructing the tangents at the points of intersection of the line and curve and finding their common point. But as we have previously shown (Art. 134), this point lies upon *u*, the polar of *U*.

(4) If any line through *U* does not cut the curve, its pole lies within the curve and is harmonically separated from every point of this line, in particular from *U*, by the curve, and hence lies upon *u*.

(6) Finally, if a line through *U* touches the curve, the point of contact lies upon *u*, and is at the same time the pole of this line.

Hence, *of two points of the plane, either each or neither lies upon the polar of the other; and of any two straight lines of the plane, either each or neither passes through the pole of the other.*

138. This and preceding theorems establish the principle of reciprocity (to which I shall return later), at least for the plane, or in general for primitive forms of two dimensions. For, to any plane figure we can construct, by the aid of a curve of the second order, a reciprocal plane figure by determining for every point of the first, its polar, and for every straight line, its pole. For this reason it will be sufficient hereafter if of two reciprocal theorems upon plane figures I demonstrate only one.

139. It was by means of the polar theory that Brianchon deduced his theorem from that of Pascal. Of each side of a hexagon inscribed in a curve of the second order he determined the pole by constructing the tangents at the vertices and finding their intersection (Fig. 43). The vertices and sides of the circumscribed hexagon correspond respectively to the sides and vertices of the inscribed hexagon; and the point of intersection of any two sides of the latter has for polar the straight line joining the corresponding vertices of the former. Since, now, the three points in which the three pairs of opposite sides of the inscribed hexagon intersect, lie in a straight line u, the three straight lines which join the three pairs of opposite vertices of the circumscribed hexagon must pass through one point U, the pole of the line u.

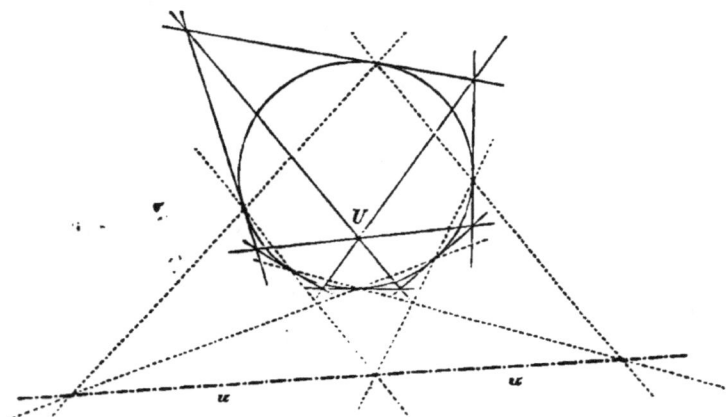

FIG. 43.

From this one application you will see what use can be made of the polar theory. With the help of a curve of the second order, for example, a sheaf of rays can be found which is reciprocal to any plane curve, each point of the curve having a ray of the sheaf as polar. To the distinctive properties of the curve there will be corresponding properties of the sheaf of rays. Later we shall have to investigate general relations of this kind.

140. If P and Q are any two points of a straight line u, their polars p and q pass through U the pole of u. Let us now choose P anywhere upon the line u, but Q at the intersection of u and p; then P, Q, and U are the vertices of a triangle of which p, q, and u are the opposite sides. Such a triangle is called a 'self-polar triangle' with respect to the curve of the second order; each vertex is the pole of the opposite side.

The following double theorem is true, as is evident without further explanation (Fig. 37):

The three pairs of opposite sides of any complete quadrangle which   *The three pairs of opposite vertices of any complete quadrilateral is inscribed in a curve of the second order intersect in the vertices of a triangle which is self-polar with respect to the curve.*

which is circumscribed to a curve of the second order lie upon the sides of a triangle which is self-polar with respect to the curve.

141. For any self-polar triangle PQU an infinite number of inscribed quadrangles can be constructed whose three pairs of opposite sides intersect in P, Q, and U. If through P we pass a secant BC to the curve (Figs. 41 and 42) and join the two curve points B and C to the point U, while BU and CU cut the curve a second time in D and A, respectively, then $ABCD$ is one of this infinite number of inscribed quadrangles, as you will easily be able to show.

If now we let the vertices A and C and the· point U of this inscribed quadrangle remain fixed while the point P moves along u the polar of U and the vertex B moves along the curve, then the straight lines PB and QB describe about the centres C and A, respectively, two projective sheaves of rays, and the points P and Q describe two projective ranges of points which are sections of these sheaves of rays. From this we observe that the sheaf of rays described about U by the polar of P, while P traverses the range of points u, is projectively related to this range of points, that is,

"If a point P traverses a range of points u, its polar p with "respect to a curve of the second order describes at the same time "a sheaf of rays U which is projectively related to this range of "points."

This theorem shows more completely the connection between any figure and its polar figure. From it we derive, for example, the following :

"If in a plane there are given two curves of the second order "k^2 and l^2, and we determine for each point of the one k^2 its "polar with respect to the other l^2, all these polars envelop a third "curve of the second order."

For if we imagine k^2 to be generated by two projective sheaves of rays U and V, to these will correspond two ranges of points projective to them and consequently to each other. These ranges of points generate the sheaf of rays of the second order which envelops this third curve.

142. In the theory of curves of the second order frequent use will be made of the following terminology :

Two points of the plane are said to be 'conjugate' with respect to a curve of the second order if one and consequently each lies upon the polar of the other.

Two straight lines of the plane are said to be 'conjugate' with respect to a curve of the second order if one and consequently each passes through the pole of the other.

Thus a point is conjugate to every point of its polar, and a straight line to every line through its pole. The vertices and likewise the sides of any self-polar triangle are conjugate, two and two; a point lying upon the curve is conjugate to itself, or is *self-conjugate*, since it lies upon its own polar, namely, the tangent at the point; and a tangent to the curve is *self-conjugate* since it passes through its own pole, namely, the point of contact.

143. If the straight line joining two points A and B, conjugate with respect to a curve of the second order, cuts the curve, then A and B are harmonically separated by the two points of intersection of the curve and straight line.

For the polar of A passes through B and contains all points which are harmonically separated from A by two points of the curve.

If from the point of intersection of two straight lines a and b, conjugate with respect to a curve of the second order, tangents may be drawn to the curve, then a and b are harmonically separated by these tangents.

For the pole of a lies upon b, and through it pass all rays which are harmonically separated from a by two tangents to the curve.

One vertex, therefore, of any self-polar triangle lies inside the curve of the second order and the other two lie outside; the curve intersects two sides of the polar triangle but not the third.

From the definition of conjugate points and straight lines it follows further that—

If two points A and B are conjugate to the same third point C, the straight line AB is the polar of C. For the polar of C must pass through both A and B.

If two straight lines a and b are conjugate to the same third line c, the point of intersection of a and b is the pole of c. For this pole must lie upon both a and b.

144. If u and v are two non-conjugate straight lines of the plane, then we know that for every point P of u there is a point P_1 of v conjugate to it; the ranges of points u and v are by this means correlated to each other projectively. For v is simply a section of the sheaf of rays U which consists of the polars of all the points of u and is projective to the range of points u (Art. 141). The straight lines joining pairs of conjugate points P and P_1 of the straight lines u and v form then a sheaf of the first or second order, according as the point of intersection of u and v is or is not self-conjugate, *i.e.* according as this point does or does not lie upon the curve. Since P is conjugate to both P_1 and U, $P_1 U$ is the polar of P, and PP_1 is conjugate to the straight line $P_1 U$. We therefore obtain this same sheaf of the first or second order if through each point P_1 of v we draw that ray which is conjugate to the straight line $P_1 U$.

If, on the other hand, U and V are two non-conjugate points of the plane, for every ray p of U there is a ray p_1 of V conjugate to it, and by correlating these the sheaves of rays U and V are projectively related to each other. For, U is then the projector of the range of points v formed of the poles of all the rays through V, and is projective to the sheaf V. The sheaves U and V therefore generate a range of points of the first or second order according as the common ray UV is or is not self-conjugate, *i.e.* according as this ray does or does not touch the given curve. Since p_1 is conjugate to both the rays p and v, the point of intersection of p and v is the pole of p_1 and the point $p_1 p$ is conjugate to pv. This same range of points, whether of the first or the second order, is obtained if in each ray p of U we determine that point which is conjugate to the intersection of p and v. Hence the theorems:

If there be given in the plane of a curve of the second order a straight line v and a point U not lying upon it, and

we determine in each ray of U that point which is conjugate to the intersection of this ray with v, all such conjugate points lie upon a curve of the second order; this curve passes through the pole V of the straight line v, through U, and through the points of contact of the two tangents, if any, which can be drawn from U to the given curve of the second order. It is only when v passes through one of these points of contact, *i.e.* when UV is tangent to the given curve,	we draw through each point of v that ray which is conjugate to the line joining this point with U, then these rays envelop a curve of the second order; this curve touches the polar u of the point U, the given line v, and the tangents at the two points in which the given curve is cut by v, if at all. It is only when U lies upon one of these two tangents, *i.e.* when u and v intersect in a point of the given curve, that we obtain a sheaf of rays of the first order instead of

that we obtain a range of points of the first order instead of a curve of the second order.*

the system of tangents to a curve of the second order.

145. Suppose now, in the theorem of the left-hand column of the last Article, that the given curve of the second order and the point U remain fixed, then to every point of the plane there will correspond one conjugate point which lies in a straight line with it through U; to every straight line of the plane there will correspond in general a curve of the second order. Similarly, if in the theorem on the right, the given curve and the straight line v remain fixed, to every ray of the plane there will correspond one conjugate ray, intersecting it in a point of v; to every sheaf of rays of the first order there will correspond in general a sheaf of rays of the second order. We obtain in this way two particular cases of the so-called 'geometric transformation of the second order.'

146. From the demonstrations in Arts. 143 and 144 we may derive the following theorems :

If a triangle AMB_1 *(Fig. 44) is inscribed in a curve of the second order, any straight line which is conjugate to one side* AB_1 *cuts the other two sides in conjugate points; and conversely, if a straight line is cut by two sides of the triangle in conjugate points it passes through the pole of the third side.*

If a triangle UVW *(Fig. 45) is circumscribed to a curve of the second order, any point which is conjugate to one vertex* W *is projected from the other two vertices by conjugate rays; and conversely, if a point is projected from two vertices of the triangle by conjugate rays it lies upon the polar of the third vertex.*

The ranges of points AM or u and B_1M or v (Fig. 44) are perspectively related to each other if to every point in u is correlated its conjugate point in v, since the common point M of u and v is self-conjugate. The centre S of the sheaf of rays generated by u and v must lie upon the tangent constructed at A, since the intersection A_1

---

* As a special case of the theorem on the left, the following may be mentioned :

"The middle points of those chords of a curve of the second order which converge toward any given actual point of the plane lie upon another curve of the second order."

The straight line v in this case lies infinitely distant, and each bisection point is harmonically separated by the curve from an infinitely distant point, and hence is its conjugate.

of this tangent with v is conjugate to the point A. Similarly S lies upon the tangent at the point B_1. Consequently S is the pole of the side AB_1, and every straight line passing through S cuts u and v in conjugate points. The theorem on the right can be proved in an analogous manner; its correctness, however, follows from the principle of reciprocity.

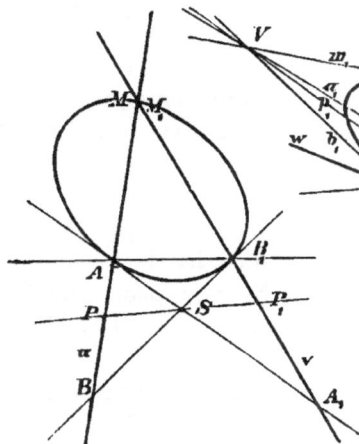 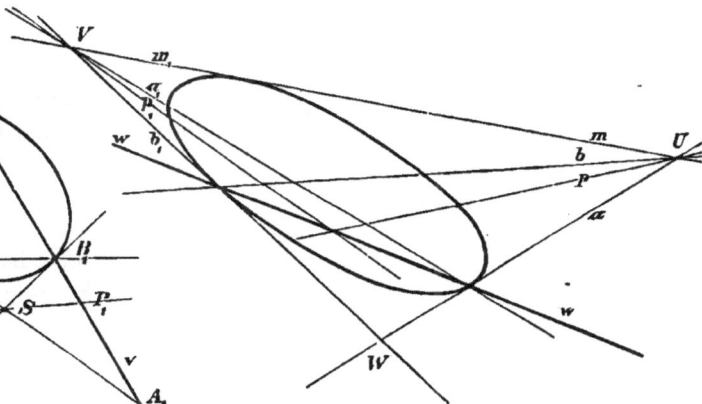

FIG. 44.   FIG. 45.

147. I shall close this series of theorems with a proof of the following one, which also admits a reciprocal:

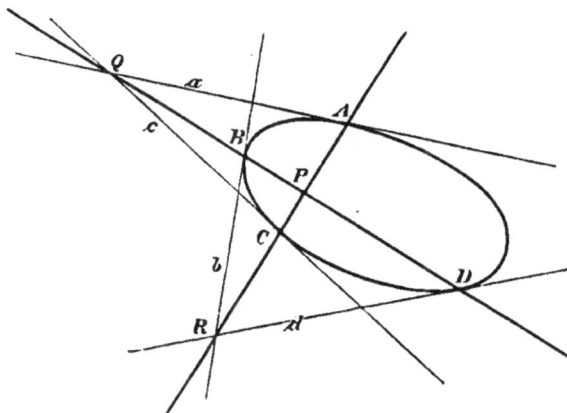

FIG. 46.

If a curve of the second order is cut by two conjugate rays AC *and* BD (*Fig.* 46), *the points of intersection* A, B, C, D, *are four harmonic points, and the tangents* a, b, c, d, *at these points are four harmonic tangents.*

The pole Q of AC, in which the tangents a and c intersect, must lie upon the straight line BD, since this is, by supposition, conjugate to AC; similarly, the point of intersection R of b and d lies upon AC. If P be the point of intersection of AC and BD, then are P, B, Q, D, four harmonic points, and RQ, b, RP, d, four harmonic rays. Thus, also, CA, CB, c, CD, are four harmonic rays, and the points ca, cb, C, cd, four harmonic points; that is, the points A, B, C, D, are projected from C, and consequently from any other point of the curve, by four harmonic rays, and the tangents a, b, c, d, are cut by c, and consequently by any other tangent, in four harmonic points.

148. All these theorems which have been enunciated for the curve of the second order may be transferred immediately to the cone of the second order, since this is cut by any plane not passing through its vertex in a curve of the second order. I shall here state only the following theorem :

"If there be given in a bundle of rays a cone of the second "order and a ray s not lying upon this cone, and through s are "passed any number of planes cutting the cone, and there are then "determined—

"1. In each plane the ray which is harmonically separated from "s by the cone;

"2. The common ray of the two planes which touch the cone "along the lines of intersection of any plane;

"3. The lines of intersection of the pairs of opposite faces of "a four-edge inscribed in the cone, whose diagonal planes are two "of the planes passed through s;

"4. The rays of contact of the two planes, if any, which can "be drawn through s tangent to the cone;

"All these rays lie in one plane σ, which is called the polar "plane of the ray s with respect to the cone."

Simply adding that s is called the 'pole-ray' of the plane σ with respect to the cone of the second order, I shall leave you to transcribe for yourselves the remaining theorems of the polar theory so as to be applicable to the cone.

EXAMPLES.

1. Construct the polar figure of any given polygon or curve lying in the plane of a curve of the second order, with respect to that curve, *i.e.* construct the polars of its points and the poles of its straight lines.

2. From any given point of a plane draw the tangents to a curve of the second order without making use of circles, *i.e.* construct the two rays of a sheaf of the second order which pass through any given point of the plane.

3. If two tangents to a curve of the second order vary so that their chord of contact envelops a second curve of the second order, their point of intersection will trace a third curve of the second order ; and conversely.

4. By linear constructions determine the polar of a given point or the pole of a given straight line with respect to a curve of the second order which is given by five conditions, *e.g.* by five points or five tangents, but which is not fully drawn.

5. If in a plane there are given two curves of the second order, any point A has a polar with respect to each of them, and if these polars intersect in A_1, A and A_1 are conjugate with respect to both curves. The polars of A_1 likewise intersect in A, and by correlating all such pairs of points as A and A_1 we obtain a one to one correspondence among the points of the plane. Similarly every straight line of the plane may be correlated to some other line of the plane to which it is conjugate with respect to both curves. In such a correlation,

The points of a straight line correspond, in general, to the points of a curve of the second order. All such curves which correspond to straight lines of the plane have in common at least one, and at most three, points, U, V, W. The two polars of every such common point coincide, so that each of them is correlated to all the points of a straight line.

The straight lines through one point correspond, in general, to the tangents to a curve of the second order. All such curves whose tangents are correlated to rays through one point have in common at least one, and at most three, tangents, u, v, w. The two poles of every such common tangent coincide, so that each of them is correlated to all the rays through one point.

If the two given curves of the second order circumscribe one quadrangle, the three pairs of opposite sides of this quadrangle intersect in the points U, V, W. These three points are the vertices, and the straight lines u, v, w, are the sides of a common self-polar triangle of the two given curves.

[If the two given curves intersect in four (real) points or in no (real) points, then, on the one hand, the curves of the second order which correspond to straight lines have three (real) points in common, and, on the other hand, the curves which correspond to points, *i.e.*, whose tangents correspond to rays through one point, have three (real) tangents in common ; the self-conjugate triangle common to the two given curves being in both of these cases wholly real. If, however, the given curves intersect in two and only two (real) points, then of the common self-conjugate triangle only one vertex and one side are real.—H.]

6. If in a bundle of rays a cone of the second order K^2 and a ray u be chosen, a one to one correspondence may be established between rays a and a_1 of the bundle which are conjugate with respect to K^2, and lie in a plane with u. If a describes a plane a, a_1 will generate a cone of the second order which passes through u and through the pole-ray of the plane a, and which has in common with K^2 any rays common to K^2 and a, and to K^2 and the polar plane of u. State the reciprocal.

[Remark : The geometrical relations expressed in examples 5 and 6 pertain to 'quadratic transformations,' of which the theory will be given more fully in the second volume of these lectures. They can be utilized in transforming simple forms into more complicated ones. Among these transformations that known as the 'Principle of Reciprocal Radii' merits a particular place, and is treated in the Appendix to this volume. It is of great importance not only in Synthetic Geometry, but also in certain investigations of Mathematical Physics and in the Theory of Functions.]

LECTURE IX.

149. From the general theory of the pole and polar we derive
the following :

" The middle points of parallel chords of a curve of the second
" order lie upon a straight line" (see Fig. 47).

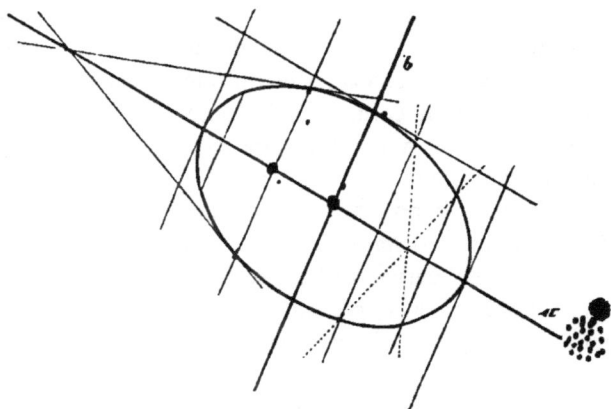

FIG. 47.

For, all these middle points are harmonically separated by the
curve from the infinitely distant point of intersection of the parallel
chords, and consequently lie upon the polar of that point.

Such a line is called a *diameter* of the curve.

The polar of any infinitely distant point of the plane with respect
to a curve of the second order is a diameter of the curve.

A diameter bisects all chords conjugate to it.

The points of contact also of the two tangents conjugate to any diameter (if such there are) lie upon this diameter. The tangents which meet the curve at the extremities of any chord conjugate to a diameter intersect in a point of the diameter (Art. 134).

150. We have already found (Art. 137) that the polars of the points of a straight line intersect in the pole of that line. For the diameters of a curve of the second order this statement becomes—

"The diameters of a curve of the second order intersect in one "point, namely, in the pole of the infinitely distant line."

If the curve is a parabola, the infinitely distant line is tangent to it, and its point of contact is its pole (Art. 135) ; hence,

"The diameters of a parabola are parallel, and pass through its "infinitely distant point."

If, on the other hand, the curve is an ellipse or hyperbola, the pole of the infinitely distant line is a finite point. This is known as the *centre* of the curve, and possesses the following property :

"Every chord of a curve of the second order which passes "through the centre is bisected in that point."

For the centre is harmonically separated from the infinitely distant point of the chord by the two points which the chord has in common with the curve.

151. The parabola has no actual centre, as is seen from the following theorem :

"If two chords of a curve of the second order bisect each other, "their point of intersection is the pole of the infinitely distant "straight line, and the chords themselves are diameters of the "curve."

The correctness of this theorem follows from the fact that this point of intersection is harmonically separated by the curve from the infinitely distant point of each chord. Since now the pole of the infinitely distant line with respect to a parabola is the infinitely distant point of the parabola, there can be no two chords of a parabola which bisect each other.

152. "The asymptotes of a hyperbola intersect at the centre."

For the tangents at the points common to a straight line and a curve of the second order always pass through the pole of the line.

The centre of a hyperbola lies outside the curve ; that of an ellipse, inside the curve (Art. 136).

H

153. To each diameter of an ellipse or hyperbola there is a conjugate diameter; of two conjugate diameters, each passes through the infinitely distant pole of the other.

" Any two conjugate diameters of an ·ellipse or hyperbola form "with the infinitely distant line a self-polar triangle ˙with respect to " the curve."

Every chord of the curve which is parallel to one of two conjugate diameters is bisected by the other, since it passes through the pole of the latter.

If one of two conjugate diameters cuts the curve,* the tangents at its points of intersection with the curve are parallel to the other diameter. Thus the conjugate to any diameter can easily be constructed.

154. *The diagonals of any paralelogram circumscribed to a curve of the second order are conjugate diameters of the curve.*

The sides of any parallelogram inscribed in a curve of the second order are parallel to a pair of conjugate diameters.

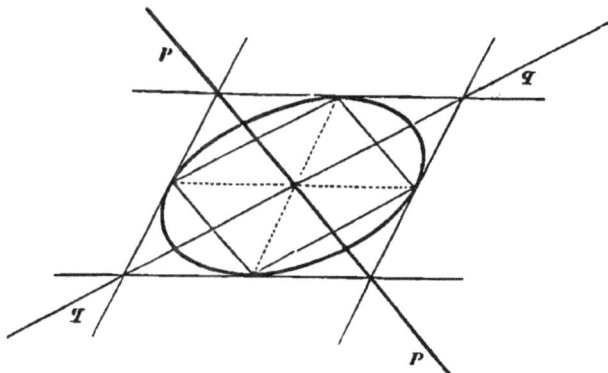

FIG. 48.

The diagonals of both the inscribed and circumscribed parallelograms (Fig. 48) are diameters of the curve, since their point of intersection in either case has the infinitely distant line for polar (Art. 134).

If we draw through ʻhis point of intersection two straight lines *p* and *q* parallel to the sides of the inscribed parallelogram, each

---

* Note that of two conjugate lines at least one must cut the curve in real points.—H.

of these bisects the sides parallel to the other ; thus p and q are conjugate diameters.

Upon p and q lie also ·the poles of the four sides of the inscribed parallelogram ; or p and q are the diagonals of that circumscribed quadrangle whose sides touch the curve at the vertices of the inscribed quadrangle. But this circumscribed quadrangle is a parallelogram, since the tangents at the extremities of a diameter are parallel, and it is evident that this circumscribed parallelogram may be considered perfectly general.

155. The second theorem of the preceding article may be stated thus:

The two chords which join any point of an ellipse or hyperbola to the extremities of a diameter are parallel to a pair of conjugate diameters.

Hence, if there be given two pairs of conjugate diameters and one point of a curve of the second order, five other points of the curve can easily be found. Draw the diameter which passes through the given point P and determine its second intersection Q with the curve, from the fact that PQ is bisected at the centre. With PQ as diagonal ˙ construct two parallelograms, the sides of each parallel to one of the given pairs of conjugate diameters ; the two new pairs of vertices of the parallelograms so determined lie upon the curve. In a similar way six tangents to a curve of the second order can easily be obtained if one tangent and two pairs of conjugate diameters are known.

156. *If all pairs of conjugate diameters of a curve of the second order are at right angles the curve is a circle.*

For under these circumstances the adjacent sides of any inscribed parallelogram are at right angles ; the parallelogram is thus a rectangle, and its diagonals are equal. That is, any two diameters of the curve, and hence all diameters, are of equal lengths.

That the circle is a curve of the second order follows from the fact

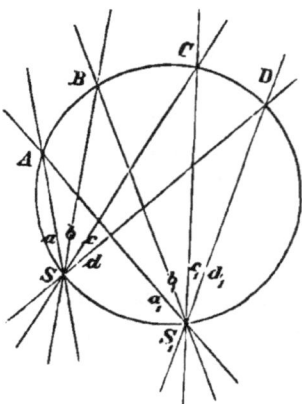

FIG. 49.

that angles at the circumference subtended by the same arc (as $\angle ASB$ and $\angle AS_1B$, Fig. 49) are equal. By virtue of this property the circle may be generated by two sheaves of rays S and S_1,

which are equal, and hence projective. In a similar way it may
be shown directly that the tangents to a circle form a sheaf of
rays of the second order. The conic section of ancient times,
namely, the curve in which a cone having a circular base is cut
by a plane, is therefore a curve of the second order, for a circular
cone is projected from any two of its rays by projective sheaves
of planes. We shall show later that not only is every conic section
a curve of the second order, but also that every curve of the second
order is a conic section, or that through any curve of the second
order cones may be passed which have circular sections.

157. Since the projective properties of circles are easily obtained,.
most authors who, like Steiner and Chasles, have based the modern
geometry upon metric relations, have chosen the circle as starting-
point for the study of curves of the second order. By choosing
this course, however, it becomes necessary to show that no other
curves of the second order than the conic sections can be generated
by projective one-dimensional primitive forms. For, if others
should exist, all theorems enunciated for the circle would need
to be investigated for them independently. For example, the
theorem that a curve of the second order is projected from any
two of its points by projective sheaves of rays, and that its
system of tangents is cut by any two tangents in projective ranges
of points, would need to be proved separately for curves of the
second order other than the conic sections. That is to say,
theorems which have been proved for the circle can be directly
extended only to sections of a circular cone.   ˙

158. *If a curve of the second order has more than one pair of
conjugate diameters at right angles, it must be a circle.*

For, if we draw from the extremities A and B of any diameter
straight lines parallel to two normal conjugate diameters, we obtain
a rectangle, which is inscribed in the curve of the second order
and is also inscriptible in a circle of which AB is a diameter.
Any second pair of normal conjugate diameters would give rise to
a second such rectangle upon the same diagonal AB. Thus the
circle would have at least four points besides A and B in common
with the curve of the second order, and would therefore wholly
coincide with it (Art. 112).

159. If two conjugate diameters are at right angles to each other
they are called *axes*, and their points of intersection with the curve
are known as the *vertices of the curve.*

The circle alone has more than one pair of axes; for it, any pair of conjugate diameters being axes.

In order to construct the axes of an ellipse or hyperbola we proceed as follows :

Construct a circle having any diameter AB of the given curve (Fig. 50) as its diameter. This in general cuts the tangents, and hence also the curve, at the extremities of this diameter. Each of the two semicircles into which the circle is divided by this diameter lies partly within and partly without the given curve, and therefore has a second intersection with it. The four points of intersection of the circle with the curve are the vertices of an inscribed rectangle whose sides are parallel to the required axes. It follows from this construction that there always exists one pair of axes for an ellipse or a hyperbola.

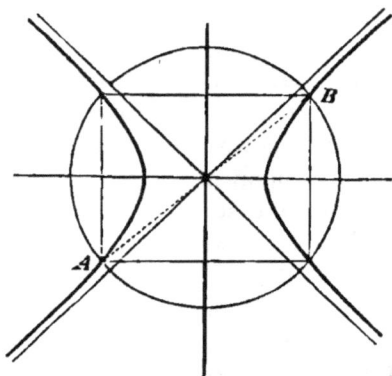

FIG. 50.

160. An axis may also be defined as a diameter of the curve which stands at right angles to the chords which are conjugate to it, and are bisected by it. The parabola has only one axis; this contains the middle points of all chords perpendicular to the common direction of the diameters.

A curve of the second order is divided into two equal symmetrical parts by each of its axes.

161. Two conjugate straight lines are harmonically separated by the tangents to the curve, which can be drawn from their point of intersection (Art. 143). Hence,

"Any two conjugate diameters of a hyperbola are harmonically "separated by the asymptotes. One of these diameters, therefore, "cuts the curve, the other does not. The axes of a hyperbola bisect "the angles between the asymptotes" (Art. 68).

162. Upon any straight line which is parallel to one of two conjugate diameters of a hyperbola, a segment is included between the asymptotes whose middle point lies upon the other diameter (Art. 68). If the straight line intersects the curve, or is tangent to the curve, the middle point of the chord, or, in the latter case, the point of contact of the line, coincides with the middle point of this segment. Hence,

"The two segments of any secant of a hyperbola which lie "between the curve and its asymptotes are equal."

"That segment of a tangent to a hyperbola which lies between "the asymptotes is bisected at the point of contact."

The first of these theorems furnishes a very simple construction for a hyperbola of which the asymptotes and one finite point are given. By it the second point of the hyperbola which lies upon any secant through the given point can be found immediately.

163. The hyperbola is intersected by but one of its axes, consequently it has two vertices; the ellipse has four vertices since it is intersected by both of its axes; the parabola has only one actual vertex, being cut by its axis a second time in the infinitely distant point.

164. A hyperbola is said to be *equilateral* if its asymptotes are at right angles to each other; the angles between any two conjugate diameters of an equilateral hyperbola are consequently bisected (Art. 68) by the asymptotes. Accordingly, if a diameter of this curve rotates about the centre, its conjugate diameter will rotate about the same point in the opposite sense, the two sheaves described by the diameters thus being equal.

The sheaves of rays by which an equilateral hyperbola is projected from the extremities of any diameter are also equal to each other; for any two corresponding rays are parallel to two conjugate diameters (Art. 155), since they intersect in a point of the hyperbola.

165. The straight line joining the middle point D of a chord AB of a parabola (Fig. 51) to the point of intersection C of the tangents constructed at A and B is a diameter of the parabola, since it is the polar of the infinitely distant point of AB (Art. 134). But C and D are harmonically separated by the two points of intersection of CD with the parabola, and one of these intersections is infinitely distant; the other, therefore, bisects the segment CD, or,

"The straight line, which joins the pole of any chord of a "parabola with the middle point of the chord, is bisected by the "parabola."

It may be shown, in a similar manner, that each of the two straight lines which can be drawn from any point of the plane to its polar, parallel to the asymptotes of a hyperbola, is bisected by the hyperbola.

166. These are some of the most important of the metric relations which are derived from the polar theory of curves of the second

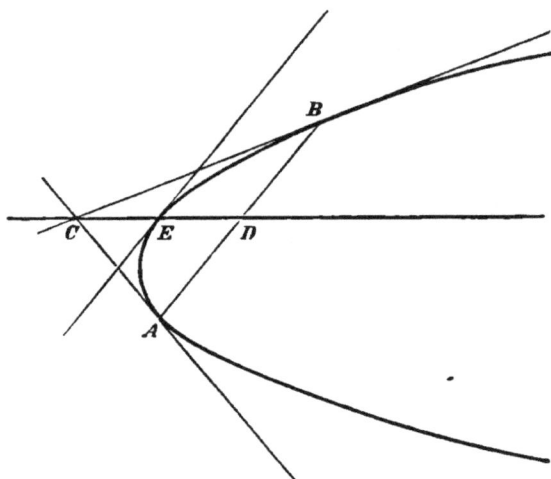

FIG. 51.

order. But from the theorems upon inscribed and circumscribed quadrangles and triangles there may be deduced other relations not wholly unimportant—in particular from the theorem of Art. 116, which may be stated as follows :

" The two diagonals of a quadrangle BB_1D_1D circumscribed "to a curve of the second order intersect in a point S, which lies in "a straight line with the points of contact of either pair of opposite "sides."

If the curve is a hyperbola, and the two pairs of opposite sides of the quadrangle are formed by the asymptotes and any other two tangents (as in Fig. 52), then S lies upon the infinitely distant straight line, and the two diagonals BD_1 and B_1D are parallel. The triangles D_1DB and D_1B_1B, which have the same base D_1B, are consequently equal in area, as are also the triangles D_1AD and B_1AB, which differ from these only by the triangle D_1AB. Therefore,

The triangles which are formed by the asymptotes of a hyperbola and a variable tangent are all of the same area.

The parallel diagonals BD_1, B_1D, and the centre A determine proportional parts upon the segments; that is,

$$AB : AD_1 = AD : AB_1 \text{ or } AB \cdot AB_1 = AD \cdot AD_1.$$

In other words, the product of the segments which a tangent BB_1 (or DD_1) makes upon the two asymptotes is constant.

Now draw through P, the point of contact of the tangent, a straight line PQ parallel to the other asymptote and meeting it in Q.

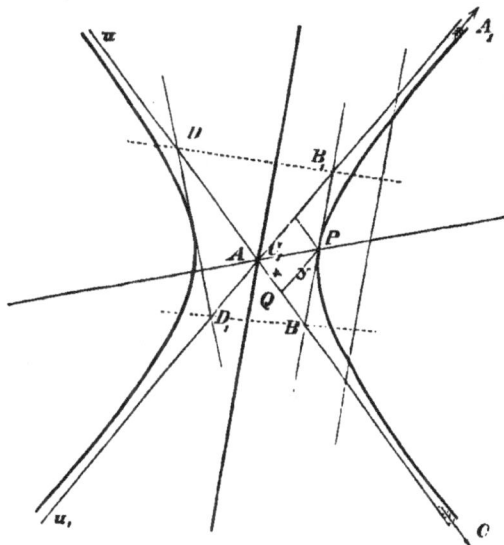

FIG. 52.

Since P bisects the segment BB_1 of the tangent (Art. 162),

$$QP \text{ or } y = \tfrac{1}{2}AB_1, \text{ and } AQ \text{ or } x = \tfrac{1}{2}AB.$$

Since now $AB \cdot AB_1$ is constant for all positions of the tangent BB_1, the product xy is constant wherever the point P may be chosen upon the hyperbola. Hence—

If we choose the asymptotes of a hyperbola for axes of a system of coördinates, the equation of that curve is

$$xy = \text{a constant.}$$

By this means the synthetic theory of the hyperbola is brought into touch with the analytical theory.

167. In the elements of analytical geometry we are accustomed to refer the ellipse and hyperbola to two conjugate diameters as

axes of coördinates, and treating our curves of the second order in the same way, we are able without difficulty to prove their identity with the analytical curves of the second order represented by equations.

Of the conjugate diameters OX and OY (Fig. 53) at least one, OX say, cuts the curve (Art. 161), and the tangents u and u_1 at the points of intersection A and C_1 are parallel to the other diameter OY. We shall first prove the following theorem :

"The product of the segments AB and C_1B_1 which any tangent "BB_1 of the curve of the second order determines upon two parallel "tangents u and u_1, is constant."

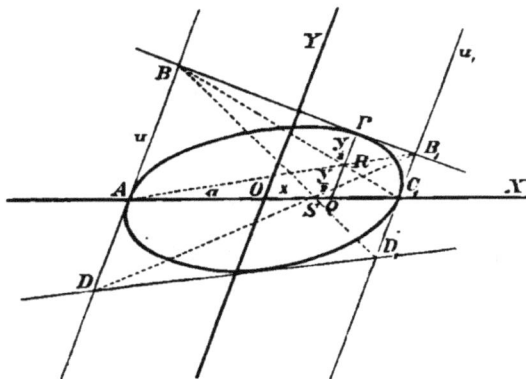

FIG. 53.

The three tangents u, u_1, and BB_1 form with any fourth tangent DD_1 a quadrangle circumscribing the curve, whose diagonals BD_1 and B_1D intersect in a point S of the diameter AC_1, and from the proportion

$$AB : C_1D_1 = AD : C_1B_1,$$

which is readily obtained, it follows that

$$AB . C_1B_1 = AD . C_1D_1,$$

and hence is constant.

This constant product is positive (equal to $+ b^2$, say) if the curve is an ellipse ; and in this case b is the semidiameter of the ellipse lying upon OY, as we readily find by drawing DD_1 parallel to OX.

In the case of the hyperbola $AB . C_1B_1$ is negative (equal to $- b^2$, say) since the segments AB and C_1B_1 have opposite senses; here b is the absolute value of the length of the segment which each of the two asymptotes determines upon the parallel tangents u and u_1.

The three tangents u, u_1, and BB_1, the latter of which touches the curve in the point P, form a circumscribed triangle, of which B, B_1 and the infinitely distant point of the diameter OY are the vertices; consequently, the three straight lines which join the vertices with the points of contact of the opposite sides intersect in one point (Art. 117).

That is, the point of intersection R of the straight lines BC_1 and B_1A lies upon the ordinate y or PQ of the point P. But since

$$\frac{QR}{AB} = \frac{QC_1}{AC_1} = \frac{PB_1}{BB_1} = \frac{RP}{AB},$$

then must $QR = RP = \frac{1}{2} QP = \frac{1}{2} y$.

We obtain the equation of the curve of the second order immediately if we multiply together corresponding sides of the equations

$$\frac{QR}{AB} = \frac{QC_1}{AC_1} \quad \text{and} \quad \frac{QR}{C_1B_1} = \frac{AQ}{AC_1},$$

and then make $QR = \frac{1}{2} y$, $AB \cdot C_1B_1 = \pm b^2$, $AC_1 = 2 \cdot AO = 2a$ and $OQ = x$, hence $QC_1 = a - x$ and $AQ = a + x$.

This gives us

$$\pm \frac{y^2}{4b^2} = \frac{a^2 - x^2}{4a^2} \quad \text{or} \quad \frac{x^2}{a^2} \pm \frac{y^2}{b^2} = 1,$$

in which the upper sign is to be used for the ellipse and the lower sign for the hyperbola. This equation is satisfied by the coördinates x, y, of any point P of the curve of the second order when it is referred to two conjugate diameters as axes.

168. If the curve is a parabola it is customary to choose as axis of ordinates an arbitrary tangent OY (Fig. 54), and as axis of abscissae the diameter OX which passes through O, the point of contact of OY. Any other two tangents AP and AP_1 which intersect OY in B and B_1, respectively, form two opposite sides of a circumscribed quadrangle of which OY and the infinitely distant line of the plane are the other two sides; consequently, the point of intersection S of the two diagonals of this quadrangle lies upon the two straight lines PP_1 and OX which join the points of contact of the two pairs of opposite sides.

Since, therefore, the quadrangle $ABSB_1$ is a parallelogram and the triangles BPS and B_1SP_1 are similar, and, moreover, the ordinates PQ and Q_1P_1, or y and y_1 of the points P and P_1, respectively, are parallel, we have

$$\frac{y}{y_1} = \frac{PS}{SP_1} = \frac{BP}{B_1S} = \frac{BP}{AB},$$

and likewise,

$$\frac{y}{y_1} = \frac{BS}{B_1P_1} = \frac{AB_1}{B_1P_1}.$$

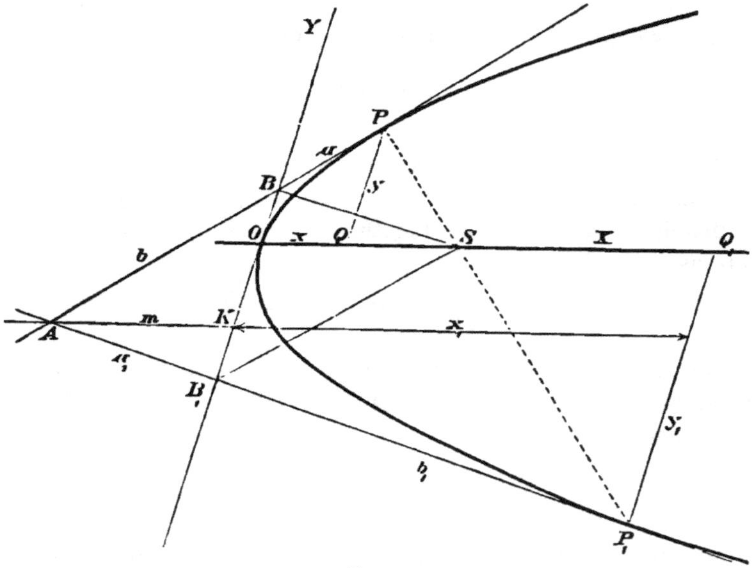

FIG. 54.

If now AK is drawn parallel to OX, meeting OY in K, we have

$$\frac{AK}{OQ_1} = \frac{AB_1}{B_1P_1} = \frac{y}{y_1}, \text{ and } \frac{OQ}{AK} = \frac{BP}{AB} = \frac{y}{y_1};$$

combine these by multiplication, and

$$\frac{OQ}{OQ_1} = \frac{y^2}{y_1^2} \text{ or } \frac{x}{x_1} = \frac{y^2}{y_1^2};$$

that is, the abscissae x and x_1 of two points P and P_1 of a parabola bear the same ratio to each other as do the squares of the ordinates. For convenience we write this equation of the parabola in the form $y^2 = 2px$, where $2p = \dfrac{y_1^2}{x_1}$.

Incidentally it appears that

$$\frac{BP}{AB} = \frac{AB_1}{B_1P_1} \quad \text{or} \quad \frac{a}{b} = \frac{a_1}{b_1},$$

and from this follows the theorem already proved, namely, that any two tangents AP and AP_1 of a parabola are cut proportionally by the remaining tangents (Art. 128).

EXAMPLES.

1. Suppose a curve of the second order is given by five conditions, *e.g.* four points and the tangent at one of them or three points and the tangents at two of them, draw any required number of diameters of the curve and find its centre.

2. Draw the chord of a given curve of the second order which is bisected at a given point.

3. Prove that the chords of a given curve of the second order which are bisected by any given chord envelop a parabola.

4. Of an ellipse or hyperbola there are given two pairs of conjugate diameters and either one point or one tangent; construct any required number of points or tangents.

5. Of a curve of the second order there are given (1) two points or two tangents and one pair of conjugate diameters, or (2) three points or three tangents and the centre ; construct the curve.

6. Construct a parabola, having given three points or three tangents and the direction of its diameters, or two points or tangents and the axis.

7. Given four tangents of a parabola ; construct its axis.

8. Construct a curve of the second order, having given three points or tangents and one axis.

9. Construct a hyperbola, having given its asymptotes and one point or one tangent.

10. In determining a curve of the second order for how many conditions (points or tangents) does each of the following parts count : (1) the centre ; (2) an axis ; (3) a pair of conjugate diameters ; (4) a self-polar triangle ; (5) a point and its polar ; (6) a pair of conjugate points or lines ; (7) a diameter ?

11. If perpendiculars be let fall from a point P upon the planes of a sheaf a, the points in which the perpendiculars meet the planes all lie upon a circle which has the perpendicular from P to a as diameter, and whose plane is normal to a. Hence—

12. If we pass through the sides a and a_1 of an oblique angle all possible pairs of normal planes, these intersect in the rays of a cone

of the second order of which a and a_1 are rays. Any plane normal to a or a_1 intersects this cone in a circle, and any plane normal to the plane aa_1 intersects it in a curve of the second order of which an axis lies in aa_1.

13. We also obtain the cone mentioned in the last example with the help of a plane a which is normal to a_1 at the point aa_1. Thus, if we let a right angle whose vertex is aa_1 so move that its plane constantly passes through the line a, while one side describes the plane a, then the other side must describe the required cone.

14. If perpendiculars are dropped from any point S upon the diameters of a curve of the second order k^2, these meet the conjugate diameters in the points of an equilateral hyperbola, which passes through S and through the centre of k^2, and whose asymptotes are parallel to the axes of k^2. The extremities of the normals to k^2 which pass through S lie upon this hyperbola (Apollonius). From no point of its plane, therefore, can more than four normals be drawn to a curve of the second order.

15. Suppose S and S_1 are two points of a curve of the second order lying upon the same diameter, and that u and u_1 are two straight lines of the plane parallel to a pair of conjugate diameters; if, now, the curve is projected from S and S_1 upon u and u_1, respectively, we obtain two similar projective ranges of points u and u_1 which are therefore proportional. A well-known simple construction for an ellipse or hyperbola is based upon this theorem.

16. If we project a parabola from one of its points upon any diameter u, and also from its infinitely distant point upon any other straight line u_1, we obtain in u and u_1 two similar projective ranges of points. From this there follows a very simple construction for a parabola.

17. In the plane of a curve of the second order k^2, a one to one correspondence may be established among lines which are perpendicular to each other and conjugate with respect to the given curve. To the rays passing through an actual point S, which does not lie upon an axis of k^2, would thus be correlated the tangents to a parabola which is touched by the polar of S and by the axes of k^2 (Example 6, p. 95). The tangents which are common to the parabola and k^2 touch the latter at the extremities of the perpendiculars drawn from S to k^2 (comp. Example 11).

LECTURE X.

THE REGULUS AND RULED SURFACE OF THE SECOND ORDER.

169. Thus far we have obtained only curves, sheaves of rays and planes, and cones of the second order, from projective one-dimensional primitive forms which either lie in the same plane or belong to the same bundle of rays. Let us now investigate whether or not other forms of the second order can be generated by means of two such projective primitive forms. In the first place we find that—

A sheaf of rays S generates with a range of points u projective to it, the same sheaf of planes of the first or second order as with the sheaf of rays by which u is projected from the centre of S. For, the plane which joins any point of u with the corresponding ray of the former sheaf passes also through the corresponding ray of the latter sheaf.

A sheaf of rays S generates with a sheaf of planes u projective to it, the same range of points of the first or second order as with the sheaf of rays in which u is cut by the plane of S. For, the point in which any plane of u is cut by the corresponding ray of the former sheaf lies also upon the corresponding ray of the latter sheaf.

Two projective sheaves of rays which lie arbitrarily in space generate (at least immediately) no new forms. For, in general, two corresponding rays will not lie in the same plane, and hence will have no point of intersection. Similarly, a range of points and a sheaf of planes projective to it generate no new forms, since a point of the former and a corresponding plane of the latter determine no third element.

170. Consequently new forms are generated only by two projective ranges of points or two projective sheaves of planes which lie arbitrarily in space. If u and u_1 are two projective ranges of points not lying in one plane, they generate a system of straight lines V, each line of which joins two homologous points of the ranges. No two straight lines of this 'regulus'* lie in one plane;

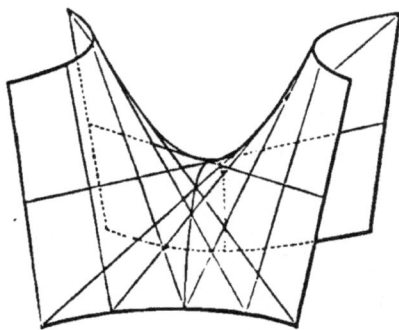

FIG. 55.   FIG. 56.

for, otherwise, two points of u and their corresponding points in u_1, and consequently u and u_1 themselves, would lie in one plane contrary to supposition. All the rays of this regulus lie upon a curved surface, called a 'ruled surface,' which is marked by the following characteristics, namely :

The surface is covered by a second system of straight lines U, each line of the one system being intersected by every line of the other system, while no two straight lines of the same system intersect each other.

| Every point which lies upon a ray of one system lies also upon a ray of the other. | Every plane which passes through a ray of one system contains also a ray of the other. |

For instance, suppose that v, v_1, v_2, are any three rays of the system V, so that each of these rays joins two corresponding points of u and u_1, and let u_2 be a straight line which also cuts these three rays v, v_1, v_2. If, then, we project the ranges of points u and u_1 from the axis u_2, we obtain two projective sheaves of planes which have the three planes u_2v, u_2v_1, u_2v_2, and consequently all their planes self-corresponding (compare Art. 84). Thus every

---

* The term 'regulus' is used by Salmon to denote a single infinity of straight lines forming a regular system—*Geom. of Three Dimensions*, 4th ed., p. 417.—H.

pair of homologous points of u and u_1 lie in a plane with u_2, and therefore u_2 is intersected by every ray of the regulus V. The same holds true of any other straight line u_3 which cuts the three rays v, v_1, v_2. Hence—

"The regulus U consists of all straight lines which cut any three "rays v, v_1, v_2, of the system V. Similarly the regulus V consists "of all straight lines which intersect three rays u, u_1, u_2, of the "system U."

Since any straight line which cuts *more than two* rays of either regulus must cut *every* ray of that regulus, each is called a 'regulus of the second order.' Any ray of one regulus is a 'director' of the other. Also either system of rays may be called the 'director system' of the other.

171. The ruled surface can thus be traversed in either of two ways by a straight line sliding along three given fixed straight lines, no two of which lie in the same plane. The three fixed lines are directors of the regulus which is described by the moving line. The moving line passes over each point of a director once, and lies once in each plane which can be passed through a director (compare Art. 39).

172. Two projective sheaves of planes u and u_1 whose axes do not lie in the same plane likewise generate a regulus V of the second order. This might be obtained from those projective ranges of points u and u_1, the first of which is a section of the sheaf u_1 by the axis u, and the second a section of the sheaf u by the axis u_1; for each straight line in which two corresponding planes of the sheaves intersect joins two corresponding points of the ranges.

173. A regulus of the second order is cut by any two of its directors in projective ranges of points.

A regulus of the second order is projected from any two of its directors by projective sheaves of planes.

Let w, w_1, and w_2 be three directors of the regulus so that every ray of the latter is intersected by these three rays. We then obtain any desired number of rays of the regulus either by passing planes through w_2 and joining the two points in which each plane is cut by w and w_1, or by choosing points in w_2 and finding the line of intersection of the planes which each point determines with w and w_1. The regulus is thus cut by the two directors w and w_1 in two ranges of points which are perspective to the sheaf of planes w_2; at the same time it is projected from

w and w_1 by two sheaves of planes which are perspective to the range of points w_2, and consequently are projective to each other.

174. Four rays of a regulus of the second order are called 'harmonic rays' if they are cut by any, and hence by every, director of the system in four harmonic points, or are projected from any director by four harmonic planes.

That is to say, if w and w_1 are two directors of this regulus, the ranges of points w and w_1 are related projectively to each other by the regulus; and if any four rays of the regulus are intersected by w in four harmonic points, their intersections with w_1 are also harmonic points. But at the same time the four rays are projected from w_1 by harmonic planes since these planes pass through four harmonic points lying in w.

175. To three arbitrary rays a, b, c, in space, of which no two lie in the same plane, a fourth ray d may be determined which is harmonically separated from one of the three by the other two. If upon any straight line which intersects the three given rays we find the fourth harmonic point to the three points of intersection, this point lies upon the required ray d. In general d is a fourth ray of the regulus of the second order to which the rays a, b, c, belong, and considered as such is harmonically separated from b by a and c.

176. If a straight line has more than two points in common with the ruled surface here considered it lies wholly upon the surface; for in that case it cuts more than two rays of one of the reguli lying upon the surface, is therefore a director of that regulus, and consequently belongs to the other regulus. On account of this property the surface is called a 'ruled surface of the second order.'

A plane which cuts the ruled surface along a ray u of one regulus lying upon it, and consequently (Art. 170) contains a ray v of the other regulus, has no point in common with the surface outside the straight lines u and v. Otherwise, every straight line of the plane passing through such a point would intersect u and v in points of the surface, and the whole plane would thus lie upon the surface, which is impossible. Since now a third line of the plane which passes through the intersection of u and v has only this point uv in common with the surface, it is tangent to the surface at this point, and we shall say that "the surface is touched by the plane of u and v in the point uv" or "the plane is tangent to the surface at this point."

I

177. "The number of planes tangent to a ruled surface of the "second order, which can be passed through a given straight line, "is equal to the number of points which the straight line has in "common with the surface. The surface is thus of the second "class."

For since in every such tangent plane there is contained a ray of one regulus lying upon the surface (and also a ray of the other), the given straight line has a point of intersection with this ray, and no two such points of intersection with rays of the same regulus can coincide, since no two rays of the same regulus lie in one plane. Hence there cannot be *more* tangent planes passed through the line than the number of its intersections with the surface. At every intersection with the surface the given line meets a ray of each regulus. The plane of the given line and either of these rays is tangent to the surface at some point along the ray; it thus contains also a ray of the other regulus, and these second rays in the two planes must intersect in a point of the given line. Hence the number of tangent planes through any line is just *equal* to the number of intersections of the surface and line. From this it follows that if through a straight line there can be passed more than two planes tangent to the surface, the line lies wholly on the surface.

178. If we consider a regulus of the second order to be generated by two projective sheaves of planes and these to be cut by an arbitrary plane σ, there appear in this plane two projective sheaves of rays such that every point of intersection of homologous rays of the sheaves lies upon a ray of the regulus.

If, on the other hand, we consider the regulus to be generated by two projective ranges of points and these to be projected from an arbitrary point S, there will arise two concentric and projective sheaves of rays such that every plane containing a pair of homologous rays of these sheaves contains also a ray of the regulus. From this follows the first part of each of the following theorems:

A regulus of the second order is intersected by any plane σ which contains none of its rays in a curve of the second order. The planes which are tangent to the ruled surface in the points of such a curve form a sheaf of the second order.	A regulus of the second order is projected from any point S which lies upon no one of its rays by a sheaf of planes of the second order. The points in which the surface is touched by the planes of such a sheaf lie upon a curve of the second order.

In order to prove the second half of the theorem on the right, we pass a plane through three of the points of contact. This cuts the sheaf of planes in a sheaf of rays of the second order, of which the three chosen points are points of contact and which envelops a curve of the second order. But this curve is identical with that in which the plane cuts the regulus, since these two · curves have the three points of contact and the tangents at these points in common. The theorem on the left is proved in an analogous way by constructing a sheaf of planes tangent to the surface, whose centre is the point of intersection of any three planes which touch the surface at points of the given curve.

179. A ruled surface of the second order is called 'a simple hyperboloid' or 'a hyperboloid of one sheet' (Fig. 55) if it contains no infinitely distant line but is intersected by the infinitely distant plane in a curve of the second order. On the other hand, it is called 'a hyperbolic paraboloid' (Fig. 56) if one and consequently (Art. 170) each regulus lying upon it contains an infinitely distant ray. / Each of the two reguli of a hyperbolic paraboloid is cut by any two of its directors in similar projective ranges of points, *i.e.* in ranges whose infinitely distant elements correspond to each other. Thus a hyperbolic paraboloid is described by a straight line which slides along two fixed straight lines u and u_1 gauche to each other (*i.e.* non-intersecting), and remains parallel to a fixed plane not parallel to either u or u_1. For, the moving line intersects not only u and u_1 but also the infinitely distant line of the given plane ; it describes therefore a ruled surface upon which lies one and consequently a second infinitely distant ray.

The hyperbolic paraboloid is cut by an arbitrary plane which contains none of its rays in a hyperbola, but when the plane is parallel to a particular straight line the section is a parabola. Any curve of section passes through the two points in which the infinitely distant rays of the surface are cut by the plane of section, and these points coincide only in case the plane contains the common point of the infinitely distant rays.

180. The hyperboloid of one sheet is not, like the hyperbolic paraboloid, touched by the infinitely distant plane, but, as we said, is cut by it in a curve of the second order. The tangent planes at the infinitely distant points of the hyperboloid are therefore actual planes which (Art. 178) intersect in one point S and form a sheaf of the second order. The cone of the second order enveloped by this sheaf converges toward the hyperboloid along its infinitely distant curve and is called its 'asymptotic cone.' An arbitrary plane which contains no ray of a hyperboloid of one sheet cuts the surface in a hyperbola, parabola, or an ellipse, according as it has in common with the infinitely distant curve of the surface two, one, or no points, or, what is the same thing, according as it is parallel to two, or to only one, or to no rays of the asymptotic cone.

181. I shall add only the following theorem, which appears directly from what has already been said:

"If straight lines be drawn through any given point parallel to "the rays of a regulus of the second order, these all lie in one "asymptotic plane or upon a cone of the second order, according "as the regulus lies upon a hyperbolic paraboloid or a hyperboloid "of one sheet."

182. A hyperbolic paraboloid is called 'equilateral' if the rays of its two reguli are parallel respectively to two planes at right angles. Each regulus of an equilateral paraboloid contains one ray which is perpendicular to a directing plane and hence to each ray of the other regulus.

EXAMPLES.

1. Show that the locus of the vertex of a cone of the second order, to which the six sides of a gauche hexagon are tangent, is a ruled surface of the second order determined by the three principal diagonals of the hexagon.

2. The three principal diagonals of a gauche hexagon whose six sides lie upon a ruled surface of the second order intersect in one point.

3. Suppose that a range of points *u* and a sheaf of rays of the first order *S*, not lying in parallel planes, are related projectively to each other; if rays be drawn through the points of *u* parallel to the corresponding rays of *S*, they will constitute one regulus of a hyperbolic paraboloid.

4. Suppose a range of points *u* and a sheaf of planes *v* are related projectively to each other, their bases not being at right angles; the perpendiculars let fall from the points of *u* to the corresponding planes of *v* form one regulus of a hyperbolic paraboloid.

5. If at the points of a straight line lying upon a ruled surface of the second order, normals are erected to the surface, these form one regulus of an equilateral hyperbolic paraboloid (Example 4).

6. If planes are passed through any chosen point normal to the rays of a regulus of the second order, these form a sheaf of the first or the second order according as the regulus belongs to a hyperbolic paraboloid or to a hyperboloid of one sheet.

7. The planes which pass through a fixed point and intersect a given hyperboloid of one sheet in parabolas envelop a cone of the second order, each ray of which is parallel to a ray of the asymptotic cone of the hyperboloid.

8. Construct a ruled surface of the second order of which there are given two rays *a* and *b* not lying in the same plane, and either three points outside *a* and *b* or three tangent planes not passing through *a* or *b*.

9. What is the locus of a point which is harmonically separated from a given point *A* by a ruled surface of the second order? What sort of intersection has this locus with a plane passing through *A*?

LECTURE XI.

183. As has been shown, five forms of the second order can be generated by projective one-dimensional primitive forms; namely, the curve or range of points of the second order, the sheaf of rays and the sheaf of planes of the second order, the cone of the second order, and the regulus of the second order. It will be convenient for us, with Von Staudt, to group these five forms of the second order and the three one-dimensional primitive forms under the common name 'elementary forms.' Among the elementary forms, then, there are two which consist of points, namely, the ranges of points of the first and the second orders; next, two which consist of planes, the sheaves of planes of the first and the second orders; and finally, four which consist of rays, namely, the sheaves of rays of the first and the second orders, the cone of the second order, and the regulus of the second order.

In the present lecture I shall undertake to show you that these elementary forms can be correlated to each other, two and two, in a manner analogous to that employed with the one-dimensional primitive forms. By so doing, the realm of our investigations is considerably enlarged; for instance, you will observe immediately that we can obtain a large number of new forms consisting of points, rays, and planes, which possess just as noteworthy properties as do those hitherto considered. At the same time we are made aware of other important theorems concerning the forms of the second order, which by this means may be obtained very easily, but otherwise with considerable difficulty.

184. Let me remind you, at the outset, of the following theorems which have been previously enunciated, and which may be fixed upon as the definitions of harmonic quadruples in forms of the second order:

Four harmonic points of a curve of the second order are projected from any fifth point of the curve by four harmonic rays (Art. 110).

Four harmonic rays of a cone of the second order are projected from any fifth ray of the cone by four harmonic planes (Art. 125).

Four harmonic planes of a sheaf of the second order are intersected by any fifth plane of the sheaf in four harmonic rays (Art. 125).

Four harmonic rays of a sheaf of the second order are cut by any fifth ray of the sheaf in four harmonic points (Art. 110).

Four harmonic rays of a regulus of the second order are cut by any director of the regulus in four harmonic points, and are projected from any director by four harmonic planes (Art. 174).

185. We may now extend the definition of the projective relation for primitive forms (given in Art. 79) so as to apply to elementary forms in general, thus—

Two elementary forms are said to be projectively related to each other if they are so correlated that any four harmonic elements of the one form correspond to four harmonic elements of the other.

It follows from this definition that two elementary forms are projective to each other as soon as they are projective to one and the same third form.

Moreover, we may extend the idea of the perspective relation between two one-dimensional primitive forms so as to apply to elementary forms in general. Thus,

Two unlike projective elementary forms are said to be in perspective position if each element of the one form lies in the corresponding element of the other.

A range of points of the second order, for example, is perspective to a cone passing through it if each ray of the latter is correlated to the point of the former which lies upon it. A range of points of the second order is projected from any one of its points by a sheaf of rays perspective to it; a sheaf of rays of the second order is cut by any one of its elements in a range of points perspective to it; a regulus of the second order is intersected by each of its directors in a range of points perspective to the regulus, etc. Two elementary forms, the one of which is derived from the other by projection or section, are obviously projectively related, since four harmonic elements of the one correspond always to four harmonic elements of the other, and when corresponding elements are superposed the forms are perspective to each other.

If to each point of a curve of the second order is correlated the tangent at this point, then is the curve related perspectively to the sheaf of rays enveloping it; for, the curve is touched in any four harmonic points by four harmonic rays of the sheaf (Art. 122). Two curves of the second order are therefore related projectively to each other if the two sheaves of rays enveloping them are projective to each other.

186. Two forms of the second order may be conveniently correlated projectively by establishing a projective relation between two one-dimensional primitive forms perspective to them. Two projective elementary forms may consequently (as in Art. 92) always be considered as the first and last of a series of elementary forms of which each is perspective to the one following.

Moreover, two elementary forms can be so correlated to each other, that three given elements of the one correspond respectively to three given elements of the other, in only one way; this has already been proved (Art. 90) for one-dimensional primitive forms, and the proof holds equally well for elementary forms in general.

For example, if it is required to correlate projectively to each other the two ranges of points of the second order, k^2 and k^2_1 (Fig. 57), which lie in one plane, so that to the points A, B, C, of k^2 correspond the points A_1, B_1, C_1, of k^2_1, we may denote by S and T_1, respectively, the points of k^2 and k^2_1, which are projected from A and A_1 by the ray AA_1, and then project the given ranges of points from the centres S and T_1 by two sheaves of rays $S(ABC...)$ and $T_1(A_1B_1C_1...)$. These are projective to the given ranges of points and consequently to each other. They are, moreover, perspective since they have the ray AA_1 self-corresponding. Any two homologous points D and D_1 of the two ranges of points are therefore projected from S and T_1, respectively, by two rays which intersect upon a fixed straight line u.

187. *If two projective elementary forms of the same kind, e.g., two ranges of points of the second order, are superposed, then all their elements are self-corresponding, or else at most two. Elementary forms which are identical are, at the same time, projective.*

Two curves of the second order which lie in the same plane and have one point S in common are correlated projectively to each other if those points of the curves are made to correspond, which lie in a straight line with S. For both curves are then perspective to the sheaf of rays S. Every common point of the curves different from S is self-corresponding; the point S is likewise self-corresponding if the curves have a common tangent in this point, *i.e.* if they touch each other at S.

Two curves of the second order which lie in the same plane and have a common tangent s are correlated projectively to each other if those tangents of the two curves are made to correspond, which intersect in a point of s. For the sheaves of the second order which envelop the curves are then perspective to the range of points s. Every common tangent of the curves different from s is self-corresponding; but s itself is self-corresponding only if the curves have a common point of contact in s.

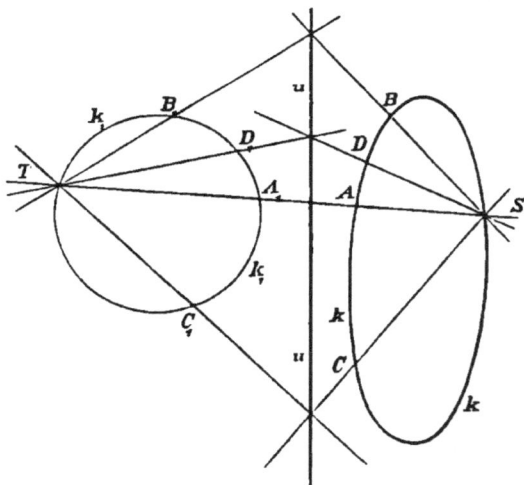

FIG. 57.

Two different curves of the second order which are correlated in the manner indicated on the left have at most three self-corresponding points. For, if they have beside S four common points, or three common points and a common tangent in S, then are they identical (Art. 112). Similarly, the two curves on the right have at most three self-corresponding tangents.

188. We are thus brought to the following reciprocal theorems:

If two projectively related curves of the second order have four self-

If two projectively related curves of the second order have four self-corresponding points, then all their points are self-corresponding and the curves are consequently identical.

corresponding tangents, then all their tangents are self-corresponding and the curves are consequently identical.

The theorem on the left may be proved as follows : The two curves can be correlated to each other in only one way so that the three points A, B, C, of the one correspond to the same three points A, B, C, of the other. But this happens if we relate the two curves perspectively to the sheaf of rays S. Suppose now that the curves have the point S also self-corresponding, then they must have a common tangent at this point and are consequently identical (Art. 112). The theorem on the right may be derived from that on the left by the application of the principle of reciprocity which we have demonstrated for the plane (Art. 138) ; however, I recommend to you as useful practice an attempt at a direct proof.

If the curves are projected from an arbitrary centre by two projectively related cones, we obtain for these wholly analogous theorems.

189. If a curve of the second order is projectively related to a regulus or to a cone of the second order, and more than three points of the curve lie upon the rays corresponding to them, then the curve is perspective to the regulus or cone ; for, it is identical with the section of the regulus or cone which lies in its plane, since it is projective to this section and has with it more than three self-corresponding points.

Similarly, a sheaf of planes of the second order is perspective to a regulus or to a sheaf of rays of the second order, which is projectively related to it, if more than three of its planes pass through the rays corresponding to them. If we project, for instance, either of the latter forms from the centre of the sheaf of planes we obtain a second sheaf of planes which is identical with the first ; for it is projectively related to the first sheaf and has with it more than three self-corresponding elements.

190. At this point the following theorems may be introduced :

Two cones of the second order which have different vertices and which are touched by the same plane along the line s joining their vertices, intersect in a curve of the second order.

Two curves of the second order which lie in different planes and which are tangent to the line of intersection s of their planes at the same point, lie upon a cone of the second order.

If each of the cones is correlated perspectively to the sheaf of planes whose axis is *s*, they will have the ray *s* self-corresponding; every other pair of homologous rays of the cones will intersect since they lie in one plane passing through *s*. The plane determined by any three points of intersection of homologous rays intersects the cones in two projectively related curves of the second order which are identical, since they have self-corresponding not only these three intersection points but also a point of *s*.

If the sheaves of rays of the second order which envelop the curves be correlated perspectively to the range of points *s*, the ray *s* will be a self-corresponding ray (compare Art. 187); every other pair of homologous rays of the sheaves determine a plane, since the rays intersect in a point of *s*. From the point of intersection of any three of these planes the two sheaves of rays may be projected by two projectively related sheaves of planes of the second order which are identical, since they have self-corresponding these three planes and also a plane passing through *s*.

The proof of these theorems may however be made much simpler. In the theorem on the left, if a plane be passed through any three points common to the two cones, the two curves of section lying in it have these three points in common and also the point of intersection with *s*; since both curves are tangent at the latter point to the straight line in which the common tangent plane of the cones is intersected, these curves of section must coincide (Art. 112). The theorem on the right may be proved in a similar way.

191. It follows incidentally that any curve of the second order may be considered as a section of a circular cone. For a circle and a given curve of the second order can be brought in an unlimited number of ways into such position that they touch each other and lie in different planes, and hence lie upon one and the same cone. *Curves of the second order are thus identical with the 'conic sections' of ancient times,* and may hereafter be designated by that name.

192. *If a sheaf of rays of the first order* S *lies in the plane of a conic section* k^2 *which is projectively related to it, but not in perspective position, then at most three rays of the sheaf pass through the points of the curve corresponding to them, and at least one.*

If a range of points of the first order lies in the plane of a sheaf of rays of the second order which is projectively related to it, but not in perspective position, then at most three points of the range lie upon the rays of the sheaf corresponding to them, and at least one.

For every sheaf of rays S_1 perspective to the curve k^2 is projective to the sheaf S, and with it generates, in general, another curve of the second order, which must have in common with the first curve every point which lies upon the ray of S corresponding to it. If more than three rays of S should pass through the points of k^2 corresponding to them, the two curves would have beside S_1 at least four common points, hence would be identical, and S would be perspective to k^2. Since every curve of the second order divides its plane into two parts, these two curves, in case they do not coincide, must either touch each other in the point S_1 or must intersect in S_1 and at least one other point P, so that each curve lies partly within and partly without the other. In the latter case the rays SP and S_1P correspond to each other, and consequently SP passes through the point P of the curve k^2 corresponding to it; in the former case, to the ray SS_1 of S, there corresponds the common tangent in S_1 and hence also the point S_1 of the curve k^2. Thus at least one point of the curve lies upon the ray of the sheaf corresponding to it.

Wholly analogous theorems are true for the forms of the first and second orders, in the bundle of rays.

193. We conclude that—

If a one-dimensional primitive form and an elementary form of the second order are related projectively to each other, and more than three elements of the one form pass through or lie upon the elements of the other which correspond to them, then the two forms are in perspective position, that is, each element of the one form passes through or lies upon the element of the other corresponding to it.

If the form of the second order is a regulus and the other either a range of points or a sheaf of planes of the first order, it can be seen immediately that these are perspective to each other if *three* rays of the regulus pass through the three corresponding points of the range or lie in the three corresponding planes of the sheaf. For the base of the range of points or axis of the sheaf of planes is then a directing ray of the regulus (Art. 170) since it intersects three rays of the latter.

194. The importance of these theorems may be judged from the following:

A sheaf of planes of the first order and a regulus or a cone of the second order projectively re-

A range of points of the first order and a regulus or a sheaf of rays of the second order projec-

lated to it generate, in general, a 'gauche curve of the third order.' This has at least one and at most three points in common with any plane.

tively related to it generate, in general, a 'sheaf of planes of the third order.' At least one and at most three planes of this sheaf pass through any point.

For, a plane cuts the regulus or the cone in a range of points of the second order perspective to it, of which in general and at most three points lie upon the corresponding planes of the sheaf.

If a range of points u of the first order and a range k^2 of the second order projectively related to it lie in one plane, the straight lines joining homologous points form a 'sheaf of rays of the third order'; at least one and at most three rays of this sheaf pass through any point of its plane.

If a sheaf of rays of the first order and one of the second order projectively related to it lie in one plane, the points of intersection of homologous rays form a 'line or curve of the third order'; this curve is intersected by any straight line of its plane in at least one and at most three points.

For if S is any sheaf of rays of the first order perspective to u, and consequently projective to k^2, at most three rays of S pass through the corresponding points of k^2, and at least one ray.

195. If the ranges of points u and k^2 of the first and second orders, respectively, have a self-corresponding point P, then every ray passing through P must be considered as a line joining two (coincident) homologous points, and the sheaf of rays of the third order includes the sheaf P of the first order as a part of it. The following theorems are therefore not to be considered exceptions to, but as particular cases of, the theorems just now proved.

If a range of points u of the first order and a range k^2 of the second order, projectively related, have two self-corresponding points A and B, they generate a sheaf of rays of the first order.

If a sheaf of rays of the first order and a sheaf of rays of the second order, projectively related, have two self-corresponding rays, they generate a range of points of the first order. ·

Suppose that to the point C of u the point C_1 of k^2 corresponds, and let S be that point of k^2 which is projected from C_1 by the ray C_1C. If now we relate u and k^2 perspectively to sheaves of rays S, these will be so related projectively to each other that to the three points A, B, C, of u will correspond the three points A, B, C_1, respectively, of k^2. But since (Art. 186) the projective relation of u and k^2 is determined uniquely by the three pairs of homologous points, the lines joining pairs of homologous points clearly form a sheaf of rays S of the first order whose centre lies upon the curve k^2.

196. A curve of the second order and two straight lines a and b each having one point in common with the curve, but which neither lie in a plane with the curve nor with each other, determine a regulus of the second order perspective to the curve and of which the two straight lines are directors.

The two sheaves of planes a and b perspective to the curve generate the regulus.

A sheaf of planes of the second order and two straight lines a and b each lying in a plane of the sheaf, but which neither intersect nor pass through the centre of the sheaf, determine a regulus of the second order perspective to the sheaf and of which the two straight lines are directors.

The two ranges of points a and b perspective to the sheaf of planes generate the regulus.

The director system of this regulus contains the rays a and b, and is likewise perspective to the curve or to the sheaf of planes, as the case may be.

If a range of points of the first order and a curve of the second order not lying in the same plane are projectively related and have a point A self-corresponding, they generate a regulus of the second order perspective to both.

If two sheaves of planes of the first and second orders respectively not belonging to the same bundle are projectively related and have one self-corresponding plane, they generate a regulus of the second order perspective to both.

In the theorem on the left, suppose that to the points A, B, C, of the range of points correspond the points A, B_1, C_1, of the curve; then the regulus determined by the curve and the two straight lines BB_1, CC_1, is perspective not only to the curve but also to the given range of points, since the three points A, B, C, of the latter lie in those rays of the regulus which correspond to them. The proof of the theorem on the right is wholly analogous.

197. From an arbitrary point not in the plane of the curve, the conic of the last article is projected by a cone of the second order, and the regulus, by a sheaf of planes of the second order. This sheaf of planes is cut by an arbitrary plane in a sheaf of rays of the second order and the regulus, in a range of points of the second order. Hence it follows :

If a range of points of the first order and a cone of the second

If a sheaf of planes of the first order and a sheaf of rays of the

order are projectively related, and one point of the former lies upon the corresponding ray of the latter, the two forms generate a sheaf of planes of the second order perspective to both.

second order are projectively related, and one plane of the former contains the corresponding ray of the latter, these two sheaves generate a range of points of the second order perspective to both.

This theorem brings us immediately to the following if we remember that any curve of the second order may be looked upon as a section of a cone of the second order :

If a range of points of the first order and a curve of the second order lying in the same plane are projectively related and have one self-corresponding point, they generate a sheaf of rays of the second order perspective to both.

If two sheaves of rays of the first and second orders, respectively, lying in the same plane are projectively related and have one self-corresponding ray, they generate a curve of the second order perspective to both.

Two projective reguli of the second order abc and $a_1b_1c_1$, of which each is the director system of the other, generate a curve of the second order and a sheaf of planes of the second order, both of which are perspective to the reguli.

The two reguli can be related projectively to each other in only one way so that to the rays a, b, c, of the one system correspond a_1, b_1, c_1, respectively, of the other system. But this happens if those two rays are correlated to each other which intersect the plane determined by the points aa_1, bb_1, cc_1, in one and the same point, or which are projected from the point determined by the planes aa_1, bb_1, cc_1, by one and the same plane.

198. *Of two projective reguli or cones of the second order, at most four pairs of homologous rays intersect unless all such pairs of rays intersect.*

If any two homologous rays in two projective reguli of the second order lie in a plane ϵ, the two systems may be projected from their directors lying in ϵ by two sheaves of planes. In case the two directors do not coincide, these sheaves of planes generate a sheaf of rays S of the first order, since they have ϵ as a self-corresponding plane. Each ray of S intersects two homologous rays of the reguli, and the reguli themselves are intersected by the plane of S in two projective curves of the second order, which have at most three self-corresponding points unless all their points are self-corresponding ; but these self-corresponding points are points of intersection of homologous rays of the reguli, and conversely. In case the two directors coincide, they cut the reguli in two projective ranges of points which have either at most two, or else all, of their points self-corresponding. At the same time the reguli are projected from these two coincident rays by two projective sheaves of planes which have either at most two, or else all, of their planes self-corresponding. In each of these planes, as well as in each of the self-corresponding points, two homologous rays of the reguli intersect. The theorem may be proved in an analogous way if for either or both of the reguli a cone of the second order be substituted.

199. From this it is clear that—

There are in general and at most four points through which pass four homologous planes in four projectively related sheaves of planes of the first order which are situated arbitrarily in space.	Of four projective ranges of points situated arbitrarily in space, there are in general and at most four sets of four homologous points which lie in one plane.

The sheaves of planes taken in pairs generate projectively related reguli or cones of the second order to which the preceding theorem is applicable. Every set of four homologous planes has a common point as soon as there exist five or more such sets.

200. Two projective curves of the second order which are superposed either generate a sheaf of rays of the second order perspective to both curves, or else there exists a point which lies in a straight line with every pair of homologous points of the curves.	Two projective sheaves of rays of the second order which are superposed either generate a curve of the second order perspective to both sheaves, or else there exists a straight line upon which every pair of homologous rays of the sheaves intersect.

Every regulus perspective to the one curve generates with its director system, which may be related perspectively to the other curve, a sheaf of planes of the second order perspective to all four forms; and according as the centre of this sheaf lies without or within the plane of the curves does the first or the second of the two cases mentioned in the theorem occur. If, then, of the straight lines joining pairs of homologous points of the curves any three pass through one and the same point U, all such lines intersect in that point (Figs. 60 and 61, p. 150).

201. Two projective curves of the second order $ABCD$ and ABC_1D_1 which have two self-corresponding points A and B, but which do not lie in the same plane, generate a form of the second order perspective to both, namely, either a regulus or a cone of the second order.

Two projectively related sheaves of planes of the second order which have two self-corresponding planes, but which are not concentric, generate a form of the second order perspective to both, namely, either a regulus or a sheaf of rays of the second order.

For, the regulus or cone perspective to the curve $ABCD$ and containing the rays CC_1 and DD_1 is perspective also to the curve ABC_1D_1 (Art. 189). The curves generate a cone if their tangents at C and C_1 intersect the straight line AB in one and the same point. Otherwise they would generate a regulus, and the plane of these tangents would also contain, besides the ray CC_1, a director of the regulus, (Art. 170), and therefore would have common with one or both of the curves a point different from either C or C_1 lying upon this director, which is impossible.

From this it follows that—

Two conics which lie in different planes, and intercept the same segment AB upon the line of intersection of these planes, can be made to lie upon either of two cones of the second order.

Two cones of the second order having different vertices, and lying in one and the same dihedral angle, intersect in one or other of two conics.

For, the conics can be correlated projectively in a twofold manner so that they have the extremities of their common chord as self-corresponding points, while the tangents at two other homologous points intersect the line AB in one point.

202. We are now prepared to prove the following theorem upon the perspective position of elementary forms of the second order:

If a curve and a sheaf of rays of the second order, or a cone and a sheaf of planes of the second order, or, in fact, any two of these four forms, are projectively related, and five elements of the one form lie in the corresponding elements of the other, then the two forms are in perspective position.

We shall choose, as the first form, a curve of the second order u^2, and for the other a sheaf of planes of the second order S^2, so situated that five points A, B, C, D, E, of the former lie in the corresponding planes α, β, γ, δ, ϵ, of the latter; all other cases can be reduced to this one. It need only be shown, then, that a rectilinear form

K

can be constructed which is perspective both to the curve and to the sheaf of planes ; for with this it will be proved that each point of the curve lies in the plane of the sheaf corresponding to it.

If the plane σ of the curve is an element of the sheaf S^2, then we obtain in it (as section of S^2) a sheaf of rays of the first order $\sigma(abcde)$ perspective to S^2, which is also perspective to the curve $\sigma(ABCDE)$, since more than three of its rays pass through the points of the curve corresponding to them (Art. 193); the point S therefore lies upon the given curve.

And conversely, if S lies upon the curve, from this point the curve is projected by a sheaf of rays of the first order $S(ABCDE)$ which is perspective to a sheaf of planes $S(\alpha\beta\gamma\delta\epsilon)$, since more than three of its rays lie in the corresponding planes of S^2; the plane u therefore is an element of the sheaf S^2.

But if the centre S does not lie upon the curve, and consequently the plane of the curve is not an element of the sheaf, let A_1 be that point of the curve which is projected from A by the plane α, that is, A_1 is the second point of intersection of α with the curve and will coincide with A only if α is tangent to the curve. Through A_1 draw, in the plane α, a straight line g different from AA_1, and with this as axis project the given curve by a sheaf of planes of the first order $g(ABCDE)$. This is projectively related to the given sheaf of planes S^2 and generates with it a regulus perspective to both sheaves (Art. 196), since α is a self-corresponding plane in the two sheaves ; this regulus is also perspective to the given curve, since four of its rays pass through the corresponding points of the curve (Art. 189). The curve u^2 and the sheaf of planes S^2 thus being both perspective to the regulus are perspective to each other.

203. If a curve of the second order $ABCDE$ and a regulus of the second order $abcde$ are projectively related to each other, but are not in perspective position, and two points A and B of the curve lie upon the corresponding rays a, b, of the regulus, these two forms generate a sheaf of planes of the second order perspective to both.

If a sheaf of planes of the second order and a regulus of the second order are projectively related to each other, but are not in perspective position, and two planes of the sheaf pass through the corresponding rays of the regulus, these two forms generate a range of points of the second order perspective to both.

The three planes, Cc, Dd, Ee, which are determined by the points

EXAMPLES.

C, D, E, of the curve and their corresponding rays c, d, e, of the regulus, intersect in one point ; the sheaf of planes which projects the regulus *abcde* from this point is perspective also to the curve *ABCDE*.

204. Two projectively related curves of the second order which lie in the same plane and have two self-corresponding points either generate a sheaf of rays of the second order perspective to both, or else there exists a point lying on neither of the two curves which is in a straight line with every pair of homologous points of the curves.

Two projectively related sheaves of rays of the second order which have two self-corresponding rays either generate a range of points of the second order perspective to both, or else there exists a straight line belonging to neither of the sheaves upon which every pair of homologous rays of the sheaves intersect.

Namely, any regulus of the second order which is perspective to the one curve generates with the other a sheaf of planes of the second order, and this in general is cut by the plane of the curves in a sheaf of rays of the second order. It is only when the centre of the sheaf of planes lies in the plane of the curves that the last case of the theorem occurs.

205. If two projectively related curves of the second order lie in one plane and have no self-corresponding points, they generate a sheaf of rays of higher order than the second, of which in general and at most four rays pass through one point. If through any point S more than four rays of the sheaf should pass, then through this point would pass an infinite number of such rays, since a sheaf of planes which has its centre at S and is perspective to one of the two curves (projects a regulus which is perspective to the curves say) is perspective also to the other.

Similarly, two projectively related sheaves of rays of the second order which lie in one plane generate a curve of higher (viz. the fourth) order, which in general has not more than four points in common with any straight line. We must, however, refrain from entering further into the discussion of these or other products of projectively related forms of the second order.

EXAMPLES.

1. To three given elements of any one of the elementary forms of the second order construct the fourth harmonic element.

2. Let S be the vertex of an angle of given magnitude which lies upon a curve of the second order and AB the chord of the curve subtended by

the angle.   If the angle be rotated about S the chord AB will generate a sheaf of rays of the second order.   In case the given angle is a right angle the sheaf of rays generated by AB is of the first order.

3. A triangle ABP is circumscribed to a curve of the second order so that its base AB lies in a given tangent to the curve and is of given length ; show that the locus of the point P is another curve of the second order.

4. If two tangents to a parabola make a constant angle with each other, show that the locus of their intersection is a hyperbola, and·that their chord of contact generates a sheaf of rays of the second order. An exception is made in case the angle between the tangents is a right angle.

5. Suppose there are given a cone of the second order and two non-intersecting straight lines a and b which are either parallel to two rays or perpendicular to two tangent planes of the cone.   If a third straight line moves so as always to intersect a and b and remain parallel to some fixed ray of the cone, or, in the second case, remain perpendicular to some fixed tangent plane of the cone, it will generate a hyperboloid of one sheet.

LECTURE XII.

THE THEORY OF INVOLUTION.

206. If two elementary forms of the same kind u and u_1, for example, two ranges of points, are projectively related and lie upon the same base, any element P of their common base may be considered as belonging either to the one form u or to the other u_1, and there correspond to it consequently two other elements, one in u_1 and the other in u. In general, these two elements corresponding to P are different from each other, as in

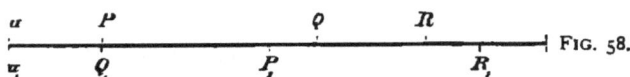

FIG. 58.

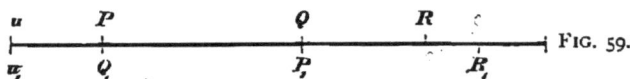

FIG. 59.

Fig. 58, where to the points P, Q, R, of u correspond P_1, Q_1, R_1, respectively, of u_1; however, it is possible for them to coincide (as in Fig. 59), so that to the element P another element P_1 corresponds *doubly*; that is, to P considered as an element of the first form u there corresponds the element P_1 of the second form u_1, and to P considered as an element of the second form u_1 there corresponds P_1 of the first u.

207. If the elements of two projective forms u and u_1 which are superposed are not all self-corresponding, but to each element another corresponds doubly, then we say that the forms have 'involutory position,' or that they are 'in involution.' Two projective forms of different kinds are said to be in involution if one of them is in involution with a section or a projector of the other.

Two projectively related ranges of points of the second order which lie upon the same conic are in involution (Figs. 60 and 61) if three and consequently (Art. 200) all straight lines joining pairs of homologous points intersect in one point; on the other hand, they are not in involution if they generate a sheaf of rays of the second order perspective to both of them. If to each point of a straight line which lies in the plane of a curve of the second order but which does not touch it, we correlate its polar with respect to the curve, we obtain a sheaf of rays which is not only projectively related to the range of points but is in involution with it (Arts. 137 and 141).

208. We can now prove the following theorem :

Two ranges of points of the second order which lie upon the same conic are in involution if to any one point A *of the curve another point* A_1 *corresponds doubly.*

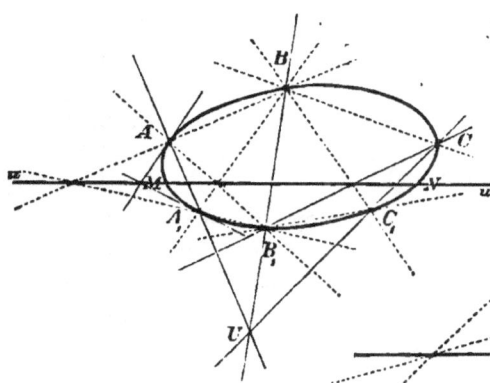
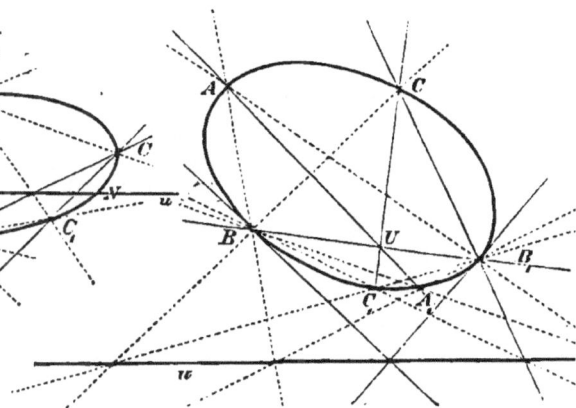

FIG. 60.   FIG. 61.

Let B and B_1 (Figs. 60 and 61) be any two homologous points of the ranges, so that to the points A, A_1, B, of the one range there are correlated the points A_1, A, B_1, of the other; let U be the point of intersection of AA_1 and BB_1 and let u be its polar with respect to the conic. The two sheaves of rays $B_1(AA_1B)$ and $B(A_1AB_1)$ which project the two ranges of points AA_1B and A_1AB_1 from B_1 and B respectively are perspective to the range of points u. For they are projective to the ranges of points of the second order and consequently to each other, and since they have a self-corresponding ray BB_1 they project one and the same range of points of the first order; but this must lie upon u since (Art. 134) the point of intersection of the homologous rays B_1A and BA_1, and likewise that of B_1A_1 and BA, lies upon this line. Any other pair of points of the conic, as C and C_1, which lie in a straight line with U are, in the same manner, projected from B and B_1 by either of two pairs of homologous rays of the sheaves perspective to u, and are therefore doubly corresponding points of the given ranges of the second order. From this and from theorems previously stated (Art. 134) it follows that—

If two projectively related curves of the second order are in involution, all straight lines joining pairs of homologous points intersect in one and the same point U, *and all points of intersection of pairs of homologous tangents lie upon the polar* u *of this point. The straight line* u *is called the 'axis of the involution,' and the point* U *the 'centre of the involution.'*

209. The two sheaves of rays of the second order by which two conics in involution are enveloped are themselves in involution, for the tangents at any two corresponding points also correspond to each other doubly. These sheaves are therefore cut by any one of their rays in two ranges of points of the first order in involution.

Likewise two conics in involution are projected from any one of their points by two sheaves of rays of the first order which are in involution, and from any point outside their plane by two cones in involution. A regulus of the second order which is perspective to the one curve is in involution with the other, and so on.

With this we may extend the preceding theorem to all elementary forms, thus :

Two elementary forms of the same kind, which are projectively related and are superposed, are in involution if any two of their elements correspond doubly.

If the two superposed elementary forms consist of rays, we may construct two coincident ranges of points of the second order perspective to them, and since these latter are in involution, two of their points corresponding doubly, the two given forms must themselves be in involution. But if the elementary forms are two sheaves of planes or ranges of points, we may construct two super-posed rectilinear forms perspective to them, and since these latter are in involution, as was just now shown, so also are the former.

210. Two forms of the same kind in involution are very frequently spoken of as a single 'involution' form, or a so-called 'involution'; the elements of this involution are said to be 'coördinated' to one another two and two, or 'conjugate' two and two, or 'paired in involution.' Thus, for example, the points of a straight line u which is chosen arbitrarily in the plane of a curve of the second order are paired in involution if every two of its points which are conjugate with respect to the curve are associated with each other (Art. 207). Similarly, the pairs of conjugate diameters of a conic form an involution sheaf. And further:

In an involution curve of the second order (Figs. 60 and 61) any two conjugate points lie in a straight line with a fixed point not lying on the curve, and any two conjugate tangents intersect upon the polar of this point with respect to the given curve.	In an involution cone of the second order any two conjugate rays lie in a plane with a fixed line not lying upon the cone, and any two conjugate tangent planes intersect upon the polar plane of this fixed line with respect to the cone.

The theorem on the left is only a repetition of that proved in Art. 208; and the theorem on the right is derived from it by projection. That is to say, if two elementary forms are in per-spective position and the elements of the one are paired in involution, so also are the elements of the other.

211. *In order to establish an involution among the elements of any elementary form, two pairs of conjugate elements A, A_1, and B, B_1, may be chosen at random; but this being done the involution is determined, and to every element of the form one and only one other is correlated.*

The involution consists, namely, of two superposed forms related projectively to each other in such a way that to the three elements A, A_1, B, of the one correspond the elements A_1, A, B_1, respectively of the other. But this relation is possible in only one way.

If the involution considered is among the elements of a range of points of the second order, the point C_1 conjugate to any fifth point C can easily be found with the help of the centre of involution U (Figs. 60 and 61), in which AA_1, BB_1, and CC_1 intersect, or with the help of the axis of involution. An analogous statement is true for the cone of the second order.

If it is required to establish an involution among the rays of a regulus of the second order we may intersect the regulus in a conic and then need only to establish an involution among the points of the conic.

The rays of a sheaf of the first order S (Fig. 62) are paired in involution by constructing a curve of the second order perspective to the sheaf, for example, a circle passing through the centre S, and then coördinating the points of this curve to one another two and two. We can proceed in a similar way with any other elementary form. An involution may, however, be established among the elements of a one-dimensional primitive form without the aid of a form of the second order.

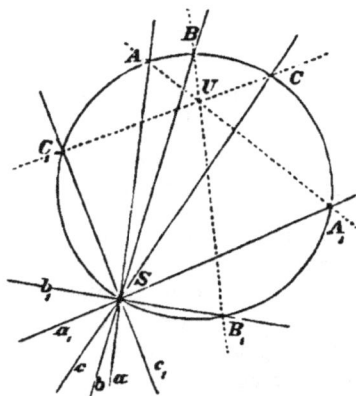

212. Two elementary forms of the same kind in involution, for example, two ranges of points of the second order (Figs. 60 and 61), are directly or oppositely projective according as two conjugate elements A and A_1 are or are not separated by two other conjugate elements B and B_1. In the first case (Fig. 61) two homologous elements, of which the one describes the range AA_1B while the other describes the range A_1AB_1, move along the base in the same sense, and can never coincide; in the latter case (Fig. 60) they move in opposite senses and must coincide twice. We shall call each self-corresponding element which occurs in two forms in involution a 'double element' or a 'focal element' of the involution.

Hence the theorem:

An involution has either two double elements or none, and is in consequence called hyperbolic or elliptic, according as two conjugate elements in it are not or are separated by two other conjugate elements. In each double element of the involution two conjugate elements coincide.

If the centre of involution U of an involution curve of the second order lies outside the curve (Fig. 60) the involution has two double points M and N, namely, the points of contact of the two tangents which can be drawn from U to the curve. The axis of involution u cuts the curve in these points since u is the polar of U (Art. 134).

213. *The double elements* M *and* N *of an involution are harmonically separated by every other pair of conjugate elements* A, A_1.

It will be sufficient if this theorem is proved for an involution curve of the second order, since every other case can be reduced to this one.

Let B, B_1, be any other pair of conjugate points (Fig. 60); then the pairs of opposite sides of the simple quadrangle ABA_1B_1 intersect in two points, conjugate with respect to the curve (Art. 142), which are harmonically separated by M and N. Thus the rays BA, BM, BA_1, BN, are harmonic since they project four harmonic points, and consequently the singular points M and N are harmonically separated by the points A and A_1. The same thing follows directly from the theorem of Art. 147, since MN and AA_1 are conjugate rays.

214. In establishing an involution among the elements of any elementary form we might choose at pleasure the two double elements M and N or one double element M and a pair of conjugate elements A and A_1; but by so doing the conjugate of every element would be fully determined. For, in the first case, any two elements of the form would be conjugate which are harmonically separated by M and N; in the second case, the other double element N could be immediately determined, since it is harmonically separated from M by A and A_1 and this case would thus be reduced to the preceding.

215. The theorems upon forms in involution which have thus far been given are of such importance, and will find such frequent application, that it seems desirable to demonstrate some of them again and in a more elementary manner, especially as new and useful theorems arise from the process. To this end we shall set out with the following definition (Art. 92):

"Two forms $ABCDE$ ... and $A_1B_1C_1D_1E_1$ ..., which are com-"posed of sets of elements in two elementary forms u and u_1, shall "be called *projective* if the forms u and u_1 can be correlated by "*projection* in such a way that to the elements A, B, C, D, E ... of "u correspond A_1, B_1, C_1, D_1, E_1 ... respectively of u_1."

We shall make use of the sign $\overline{\wedge}$ to denote projectivity. For example, if u and u_1 are two ranges of points of the first order which lie in the same plane, but not upon the same straight line, then $ABCDE ... \overline{\wedge} A_1B_1C_1D_1E_1 ...$
only if the straight lines AA_1, BB_1, CC_1, DD_1, EE_1 ... pass through

one and the same point or are tangent to a conic, to which u and u_1 are also tangent.*

A set of four elements ABCD *in definite order, chosen arbitrarily from an elementary form* (a '*throw*'), *is projective to every permutation of these elements in which one pair and also the other pair are interchanged.*

That is,
$$ABCD \ \overline{\wedge}\ BADC \ \overline{\wedge}\ CDAB \ \overline{\wedge}\ DCBA.$$

Suppose, for example, $ABCD$ is a 'throw' upon a straight line, all other cases being reducible to this one, and let it be re-quired to show that

$$ABCD\ \overline{\wedge}\ CDAB.$$

Project $ABCD$ from an arbitrary point S (Fig. 63) upon a straight line passing through A, denoting the projection by

$$AEFG.$$

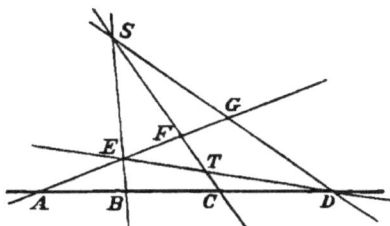
FIG. 63.

Let T be the intersection of CF and DE.

Then,

$ABCD$ is the projection of $AEFG$ from the centre S,

$AEFG$ „ „ $CTFS$ „ „ D,

$CTFS$ „ „ $CDAB$ „ „ E.

Hence, $\quad ABCD\ \overline{\wedge}\ AEFG\ \overline{\wedge}\ CTFS\ \overline{\wedge}\ CDAB$,

and consequently, $\quad ABCD\ \overline{\wedge}\ CDAB$

The other relations may be proved in a similar way. We infer from this that

"If $abcd\ \overline{\wedge}\ ABCD$ then also $abcd\ \overline{\wedge}\ BADC\ \overline{\wedge}\ CDAB\ \overline{\wedge}\ DCBA$."†

---

* The relation $ABCD\ \overline{\wedge}\ A_1B_1C_1D_1$ also signifies, as has already been shown (Art. 97), that among the segments of the straight line u and u_1 there exists the following proportion : $\dfrac{AB}{AD} : \dfrac{CB}{CD} = \dfrac{A_1B_1}{A_1D_1} : \dfrac{C_1B_1}{C_1D_1}.$

† With the help of this important theorem the following among other remarkable relations may be proved :

The six vertices of any two self-polar triangles of a conic k^2 lie upon a second conic, to which an infinite number of self-polar triangles of the first can be inscribed.

The six sides of any two self-polar triangles of a conic touch a second conic, to which an infinite number of self-polar triangles of the first can be circumscribed.

For suppose ABC and DEF to be the two self-polar triangles, no three

216. The theorem of Art. 209, already proved in a different manner, viz., "two superposed elementary forms which are pro-"jectively related are in involution if any two elements A and A_1 "correspond doubly," results directly from the relation proved in the last article. For, if to any element B of the one form the element B_1 of the other corresponds, so that to the elements A, A_1, B, of the former correspond the elements A_1, A, B_1, respectively of the latter, it follows, since

$$AA_1BB_1 \barwedge A_1AB_1B$$

by virtue of permutation, that to the element B_1 of the first form the element B of the second corresponds, or that any two elements B and B_1 correspond to each other doubly.

A consequence which might have been stated earlier is this:

"A range of points of the first order u and a sheaf of rays S "projective to it are in involution if the centre of the sheaf lies "outside u, and, of two points P and P_1 of the range, each lies upon "that ray of the sheaf which corresponds to the other."

For the section of the sheaf of rays made by the straight line u is projective to the range of points upon u and is in involution with it, since the points P and P_1 correspond to each other doubly. In a similar way we determine when a sheaf of planes is in involution with a range of points or with a sheaf of rays.

217. The fact that any two conjugate elements A, A_1, of an involution are harmonically separated by the double elements M and N, if such appear, may also be proved in an elementary way. Let the involution $MNAA_1$ consist of two superposed projective ranges of points of the first order. Then to the points M, A, N, A_1, of the one range correspond the points M, A_1, N, A, of the other,

---

of whose six vertices lie in one straight line. Then the sheaves of rays $A(BCEF)$ and $D(CBFE)$ are projectively related (Art. 144), since to the four rays AB, AC, AE, AF, of the first sheaf, the rays DC, DB, DF, DE, of the second are conjugate with respect to the conic k^2. But from the relation $A(BCEF) \barwedge D(CBFE)$ it follows that $A(BCEF) \barwedge D(BCEF)$ and therefore the six points A, B, C, D, E, F, lie upon a curve of the second order as the theorem on the left asserts. If now D' and E' are two points of this curve which are conjugate with respect to the first-named conic k^2, and consequently vertices of a self-polar triangle $D'E'F'$, the third vertex F' lies also upon the conic through A, B, C, D, E, F; for the conic passing through A, B, C, D', E', F', has five points in common with that through A, B, C, D, E, F, and hence coincides with it.

The theorem on the right is proved in a similar way.

since the double points M and N are self-corresponding, and A and A_1 correspond doubly. That is,

$$MANA_1 \barwedge MA_1NA.$$

If now we project $MANA_1$ from an arbitrary point S (Fig. 64) upon a straight line passing through M, and denote the projection by $MRKT$, then this is projectively related to $MANA_1$, and consequently to MA_1NA.

But these projective forms have the point M self-corresponding and consequently are in perspective position, that is, the straight lines RA_1, KN, TA, intersect in one and the same point Q. By this means

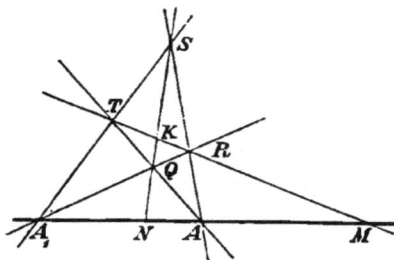

FIG. 64.

we obtain a quadrangle $QRST$ of which two pairs of opposite sides intersect in A and A_1, while the diagonals pass through M and N respectively; thus the points $MANA_1$ are clearly harmonic points.

218. A group of three pairs of elements $AA_1 . BB_1 . CC_1$ of an involution is sometimes called an involution in the original and strict sense of the word. These six elements are not independent, since in an involution the conjugate to any element is determined as soon as two pairs of conjugate elements are given, and since any two forms which are composed of elements chosen from these six symmetrically, as, for instance, AA_1BC and $A_1AB_1C_1$, or AB_1C_1C and A_1BCC_1 are projectively related.

Conversely, it follows from the relation $AA_1BC \barwedge A_1AB_1C_1$ that the three pairs of elements A, A_1; B, B_1; C, C_1 form an involution $AA_1 . BB_1 . CC$; for, in the projective forms AA_1BC and $A_1AB_1C_1$, the elements A and A_1 correspond doubly, and consequently the elements B, B_1, and C, C_1, must correspond doubly.

A double element M or N can take the place of a pair of elements in the involution; thus, for example, $M . AA_1 . BB_1$ is an involution if $MAA_1B \barwedge MA_1AB_1$. Similarly, $M . N . AA_1$ is an involution if $MANA_1$ is projective to MA_1NA, the four elements thus forming a harmonic quadruple.

219. We can now prove the following theorems:

The three pairs of opposite sides of a complete quadrangle $QRST$ (Fig. 65) are cut by any straight line which lies in the plane of the quadrangle and passes through none of its vertices, in three pairs of points in involution.

The three pairs of opposite vertices of a complete quadrilateral are projected from any point which lies in the plane of the quadrilateral but upon none of its sides, by three pairs of rays in involution.

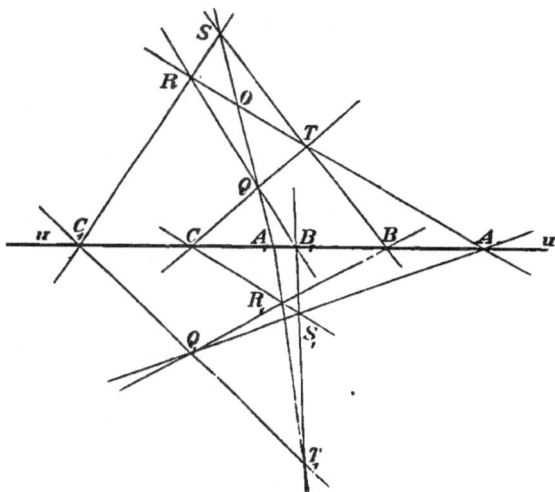

FIG. 65.

Let u (Fig. 65) cut the sides RT and SQ, ST and QR, QT and RS, in the points A and A_1, B and B_1, C and C_1, respectively, and let O be the point of intersection of QS and RT. Then $ATOR$ is a projection both of ACA_1B_1 from the centre Q and of ABA_1C_1 from the centre S, and therefore,
$$ACA_1B_1 \barwedge ATOR \barwedge ABA_1C_1.$$
Moreover, by interchanging A with A_1 and B with C_1 we obtain (Art. 215)
$$ABA_1C_1 \barwedge A_1C_1AB.$$
Therefore
$$ACA_1B_1 \barwedge A_1C_1AB,$$
and hence $AA_1 \cdot BB_1 \cdot CC_1$ is an involution.

For, if two ranges of points lying in u are so related projectively that to the points A, C, A_1, of the first correspond A_1, C_1, A, respectively, of the second, the two forms are in involution since A and A_1 correspond doubly; B and B_1 are conjugate elements of the involution on account of the relation $ACA_1B_1 \barwedge A_1C_1AB$.

If, now, of a range of points u in involution, two pairs of points A, A_1, and B, B_1, are given and it is required to find the point C_1 conjugate to any fifth point C, we can proceed without making use of the form of the second order in the following manner. Construct a complete quadrangle of which two opposite sides pass through A and A_1 respectively, two others through B and B_1 respectively, and a fifth side through C; then the sixth side will intersect the given line in the required point C_1.

Two quadrangles $QRST$ and $Q_1R_1S_1T_1$ constructed as above have essentially different positions with respect to the involution $AA_1 . BB_1 . CC_1$ if the sides passing through A, B, C, in the one quadrangle intersect in a point T, while in the other they form a triangle $Q_1R_1S_1$, their sixth side in both cases passing through C_1 (Fig. 65). If the two quadrangles lie in different planes their eight vertices form two tetrahedra $QRST_1$ and $Q_1R_1S_1T$ which are both inscribed and circumscribed to each other. If the one tetrahedron is given the other can be easily constructed in an infinite number of ways.

If two points M and N of the straight line u are harmonically separated by A and A_1, and also by B and B_1, that is, by two pairs of opposite sides of the quadrangle $QRST$, then are they also harmonically separated by C and C_1, that is, by the third pair of opposite sides of the quadrangle. For, in this case, M and N are the double points of the involution $AA_1 . BB_1 . CC_1$.

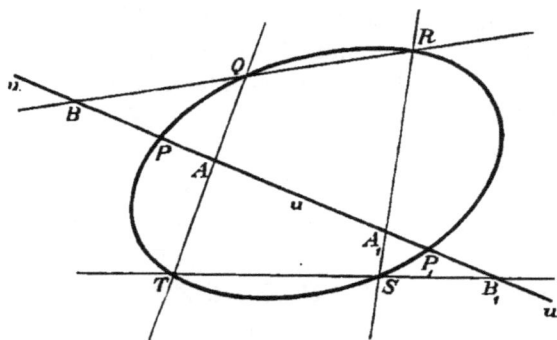

FIG. 66.

220. "If a curve of the second order is circumscribed about "a simple quadrangle $QRST$ (Fig. 66) the three pairs of points "in which any straight line u passing through no vertex of the

"quadrangle is cut by the curve and by the two pairs of opposite "sides of the quadrangle form an involution" (Desargues' theorem).

Suppose the chosen line u is cut by the sides TQ, QR, RS, and ST in the points A, B, A_1, B_1, respectively, and by the curve in the points P and P_1. Then the two sheaves of rays by which the points P, R, P_1, T, of the curve are projected from Q and S are projectively related, and consequently the ranges PBP_1A and $PA_1P_1B_1$ in which these sheaves are cut by the straight line u are also projectively related. Since now $PA_1P_1B_1 \barwedge P_1B_1PA_1$ (Art. 215) it follows that

$$PBP_1A \barwedge P_1B_1PA_1$$

i.e. $PP_1 . AA_1 . BB_1$ is an involution.

A wholly analogous theorem holds for the curve of the second order which is inscribed in a simple quadrilateral.

221. Since an involution in a one-dimensional form is completely determined by two pairs of elements A, A_1, and B, B_1, we have the following theorems:

The curves of the second order circumscribed about any fixed quadrangle are intersected by a straight line u which lies in the plane of the quadrangle, but which passes through none of its vertices, in point-pairs of an involution. The straight line is touched by two of these curves in the double points of the involution.	The pairs of tangents to the curves of the second order inscribed in any fixed quadrilateral, which pass through one point S lying in the plane of the quadrilateral but upon no one of its sides, form an involution sheaf. The double rays of this involution touch two of the conics in S.

There are either two or no curves of the second order which are

circumscribed to the quadrangle and at the same time are touched by the given straight line,	inscribed to a quadrilateral and at the same time pass through the given point,

according as these double elements do or do not appear.

Thus the problem to construct the conics which pass through four given points and touch a given line, or which touch four given lines and pass through a given point, is reduced to finding the double elements of an involution. We shall be concerned with this problem of the second order in a subsequent lecture.

If, in the left hand theorem above, the straight line u lies infinitely distant, this particular case presents itself:
"Through four actual points of a plane there can be passed "either two or no parabolas."

EXAMPLES.

1. If an involution be chosen upon a tangent to a conic section and from the pairs of conjugate points new tangents to the conic be drawn, all such pairs of tangents will intersect upon a definite straight line.

2. The vertices of all right angles whose sides are tangent to a given parabola lie upon a fixed straight line. Also, the vertices of all isosceles triangles whose bases lie upon a fixed tangent to a parabola and whose sides touch the parabola, lie upon a fixed straight line.

3. The vertices of all triangles circumscribed to a conic, whose bases remain on a fixed tangent and are bisected by its point of contact A, lie upon the diameter of the curve passing through A. The straight lines which in the several triangles join the points of contact of the other two sides are parallel to the base.

4. If two planes rotate about two fixed straight lines in such a way as always to be parallel respectively to conjugate diameters of a given conic, their line of intersection will describe a ruled surface, or else a cone of the second order, upon which the two fixed lines lie.

5. Two projectively related sheaves of rays or sheaves of planes being given, how can they be brought into such a position as to form an involution?

6. Of an involution upon a straight line there are given two pairs of conjugate points A, A_1, and B, B_1. Construct by means of the complete quadrangle the conjugate C_1 to any fifth point C; and, in particular, determine the 'centre' of the involution, *i.e.* the point whose conjugate lies infinitely distant.

7. Two pairs of conjugate diameters of a conic section are known, draw the conjugate to any fifth diameter.

8. If through any point straight lines be drawn parallel to the three pairs of opposite sides of a complete quadrangle they form an involution. Hence:

9. If two pairs of opposite sides of a complete quadrangle intersect at right angles, so also does the third pair. Or, the three perpendiculars let fall from the vertices of a triangle upon the opposite sides intersect in one point, the so-called 'orthocentre' of the triangle.

10. All curves of the second order which pass through the three vertices and the orthocentre of a triangle are equilateral hyperbolas.

L

An equilateral hyperbola can always be circumscribed to any quadrangle : this passes through the orthocentres of the four triangles formed by the vertices of the quadrangle.

11. The sides of any triangle form with the infinitely distant line of the plane a complete quadrilateral whose three pairs of opposite vertices are projected from the orthocentre of the triangle by three pairs of rays at right angles.

12. The two tangents to a parabola which can be drawn from the orthocentre of a circumscribed triangle intersect at right angles. Consequently the orthocentres of all triangles circumscribed to a parabola lie upon a fixed straight line. (Ex. 2.) In particular, the orthocentres of the four triangles formed by the sides of any complete quadrilateral lie upon one straight line.

13. If a rectangle $ABCD$ is circumscribed to an ellipse or a hyperbola, the tangents which can be drawn from any point S of the circle through the vertices A, B, C, D, intersect at right angles ; for the two pairs of opposite vertices of the rectangle are projected from S by two pairs of normal rays.   Hence we conclude that "the vertices of all right angles whose sides are tangent to an ellipse or hyperbola lie upon a circle." This might be looked upon as a special case of the following :

14. Any two tangents to a conic section k^2 which pass through two conjugate points of a given involution range u intersect, in general, upon another fixed conic.	Any two points of a conic section k^2 which lie upon two conjugate rays of a given involution sheaf S, are in general upon the same ray of a sheaf of the second order.*

An exception arises if the curve k^2 is touched by the straight line u (Ex. 1).   In general the conic can be circumscribed by a quadrilateral $abcd$ of which two opposite vertices ac and bd coincide with two conjugate points P and P_1 of the involution u. If now two tangents to k^2 which pass through two other conjugate points Q and Q_1 of u intersect in a point S, then these tangents SQ and SQ_1 together with the rays SP and SP_1, and with that third pair of rays by which two opposite vertices R and R_1 of the quadrilateral, different from P and P_1, are projected from S form an involution (Art. 220), and consequently SR and SR_1 pass through two conjugate points of the involution u. If then we correlate the sheaves R and R_1 to each other projectively so that homologous rays pass through conjugate points of u, they will generate a curve of the second order which is the locus of the point S.

---

* This sheaf of the second order degenerates into two sheaves of the first order in case the conic is touched by two conjugate rays of the involution S.

The theorem on the left might be stated : "The pairs of tangents to a curve of the second order which are harmonically separated by two given points intersect in general upon another curve of the second order." Thus expressed it becomes a special case of the following, the proof of which is left to the student :

| The pairs of tangents to a curve of the second order, which are conjugate with respect to a second curve of the second order, intersect in general in the points of a third curve of the second order. | The pairs of points of a curve of the second order, which are conjugate with respect to a second curve of the second order, lie in general upon the tangents to a third curve of the second order. |

It is easily seen that the third curve (on the left) passes through the points of contact of each of the tangents common to the first two.

15. If, of the three circles which have the diagonals of a complete quadrilateral for diameters, any two intersect, the third passes through their points of intersection. The sides of any right angle which has its vertex at one of these points of intersection touch a curve of the second order inscribed to the quadrilateral.

16. All hyperbolas circumscribing a quadrangle whose vertices lie upon a circle have parallel axes ; the directions of these axes bisect the angles which the pairs of opposite sides make with each other. For, the double rays of an involution of which three pairs of conjugate rays are parallel to the three pairs of opposite sides of the quadrangle are at right angles to each other.

17. If through the middle point M of a given chord of a curve of the second order two secants of the curve be drawn, these determine an inscribed quadrangle whose other two pairs of opposite sides intercept segments of the given chord which are likewise bisected at the point M.

LECTURE XIII.

METRIC PROPERTIES OF INVOLUTIONS. FOCI OF CURVES
OF THE SECOND ORDER.*

222. Let A, A_1 ; B, B_1 ; C, C_1 (Figs. 67 and 68), be three pairs of points of a range in involution, so that (Art. 219) among others.

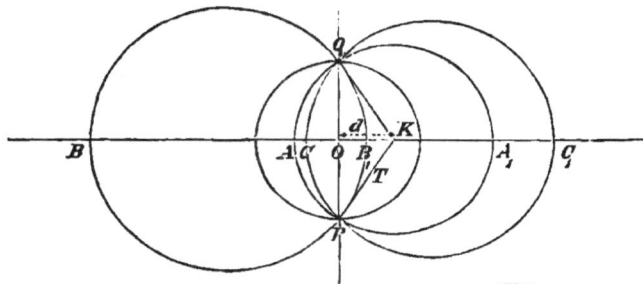

FIG. 67.

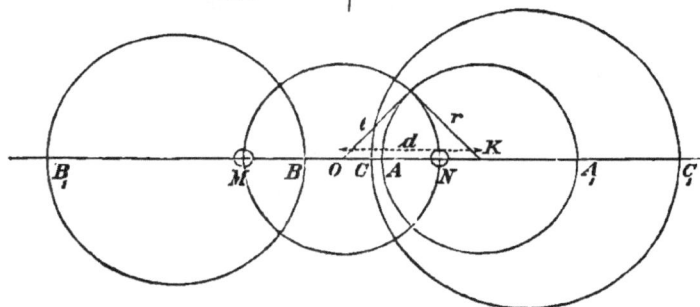

FIG. 68.

we have the relation $AA_1BC_1 \barwedge A_1AB_1C$; then, for the segments

---

* The more important focal properties of conic sections, aside from those relating to the directrices and the fundamental properties which we use as the definition of foci, were known to Apollonius. (Compare *Conicorum*, Liber III., Prop. 45. *et sequitur.*)

which the six points determine upon the line we have the following proportion, viz. :

$$\frac{AA_1}{AC_1} : \frac{BA_1}{BC_1} = \frac{A_1A}{A_1C} : \frac{B_1A}{B_1C}.$$

Instead of this we may write :

$$\frac{AA_1}{AC_1} : \frac{BA_1}{BC_1} = \frac{AA_1}{CA_1} : \frac{AB_1}{CB_1} ;$$

for $A_1A = -AA_1$; $A_1C = -CA_1$, etc., since two such segments are of equal length, but are taken in opposite senses. If all denominators and the common factor AA_1 are removed, we have the equation

$$AB_1 . BC_1 . CA_1 = AC_1 . BA_1 . CB_1. \quad\ldots\ldots\ldots\ldots(1)$$

This relation is found incidentally in the *Collections of Pappus,* Liber VII., Prop. 130.

The relation $AA_1BC_1 \barwedge A_1AB_1C$ does not lose its validity if two conjugate points are interchanged. The same thing is true therefore in equation (1).

By interchanging C and C_1, for example, we obtain the equation

$$AB_1 . BC . C_1A_1 = AC . BA_1 . C_1B_1, \quad\ldots\ldots\ldots\ldots(1a)$$

and by making different interchanges other similarly constructed equations would result from (1). Wholly analogous equations may be expressed for the sines of the angles which six elements of an involution sheaf of rays of the first order form with one another.

223. Equations (1) and (1a) assume a much simpler form if one of the six points (for example, C_1) becomes infinitely distant. Its conjugate point C in this case becomes coincident with the so-called centre of the involution, that is, with that point O which is co-ördinated to the infinitely distant point.

At the same time the ratios

$$\frac{BC_1}{AC_1} = \frac{AC_1 - AB}{AC_1} = 1 - \frac{AB}{AC_1}$$

and

$$\frac{C_1A_1}{C_1B_1} = \frac{C_1B_1 - A_1B_1}{C_1B_1} = 1 - \frac{A_1B_1}{C_1B_1}$$

approach indefinitely near to the value unity, since AC_1 and C_1B_1 increase indefinitely, while AB and A_1B_1 are finite segments. Equations (1) and (1a) thus take the form

$$AB_1 . OA_1 = BA_1 . OB_1$$

and

$$AB_1 . BO = AO \ BA_1$$

respectively.   Dividing the first of these by the second we obtain

$$\frac{OA_1}{BO} = \frac{OB_1}{AO} \; ;$$

or $OA \cdot OA_1 = OB \cdot OB_1.$ ....(2)

That is, " The product of the two segments determined by the
" centre and any two conjugate points of an involution range of the
" first order is constant."

If the range has two double points M and N (Fig. 68), in each
of which two conjugate points coincide, it follows from (2) that

$$OA \cdot OA_1 = OM^2 = ON^2. \quad ....(3)$$

The centre O thus bisects the segment MN between the double
points.   At the same time this equation expresses the fact (Art. 72)
that M and N are harmonically separated by A and A_1, as we
already know.

According as the product $OA \cdot OA_1$ is positive (equal to a square
OM^2) or negative, do the double points appear or not.   In the
former case A and A_1 lie upon the same side of the centre O, in
the latter upon opposite sides.   This same conclusion is reached
in a previous theorem (Art. 212).

224. If a circle be described upon each segment AA_1, BB_1, CC_1,
etc., which is determined by a pair of conjugate points of an involution, as diameter, and the radius of any one of these circles,
say that upon AA_1, be denoted by r, while d denotes the distance
of its centre from O, then (Figs. 67 and 68)

$$OA = OK - AK = d - r$$
and $$OA_1 = OK + KA_1 = d + r \; ;$$
consequently,

$$OA \cdot OA_1 = (d - r)(d + r) = d^2 - r^2.$$

If the involution has double points M, N (Fig. 68), that is, is
hyperbolic, then O lies outside the circles and $d^2 - r^2$ is equal to
the square on the tangent t, which can be drawn from O to the
circle ; for t and r are the sides of a right-angled triangle, of which
d is the hypotenuse.

The length of this tangent is, from equation (3), the same for
all circles constructed upon the segments, and is equal to half
the length of the segment MN between the double points of the
involution.

If, then, a circle is described upon MN with centre O, it cuts all circles described upon AA_1, BB_1, CC_1, etc., at right angles.

It may be shown, moreover, that the latter circles are cut orthogonally by every circle passing through M and N.

When the involution is elliptic, *i.e.* has no (real) double points (Fig. 67), its centre O lies within the circle described upon AA_1, and $d^2 - r^2$ is negative.

If now through O, at right angles to AA_1, a chord PQ of this circle be drawn, either half of it, OP or OQ, forms one side of a right-angled triangle of which d is the other side and r the hypotenuse, so that $d^2 + OP^2 = r^2$, or $d^2 - r^2 = - OP^2 = - OQ^2$. But from equation (2) of the last article the expression $d^2 - r^2$ is constant, hence the length of this half chord remains the same for all circles constructed upon the segments AA_1, BB_1, CC_1, etc., and consequently, these circles all pass through the two points P and Q. The angles APA_1, BPB_1, CPC_1, etc., are therefore right angles, and we obtain the theorem :

If an involution range of the first order has no double points, in any plane containing the range there are two points P *and* Q, *from which it may be projected by a sheaf of rays in involution, such that any two conjugate rays of the sheaf are at right angles to each other.*

That the circles described upon AA_1, BB_1, CC_1, etc., all pass through the points of intersection of any two of them may also be concluded from the following statement :

" An involution in a sheaf of rays is rectangular if any two of its "rays a and b are at right angles to their conjugates a_1 and b_1."

The correctness of this assertion follows from the fact that the rays of a sheaf S can be paired in involution in only one way so that its rays a and a_1 as well as b and b_1 are conjugates. But this happens if to each ray is coördinated the ray at right angles to it.

225. At this point we may insert the following theorem :

If a right-angled triangle inscribed in a curve of the second order be permitted to vary in any manner so that its vertex remains fixed, its hypotenuse will constantly pass through a fixed point.

The points of the curve are paired in involution by the rays of the rectangular involution sheaf S_1 (comp. Art. 210).

226. If in the plane of a curve of the second order k^2, a sheaf of rays of the first order U whose centre does not lie upon the curve is given, the rays of U are paired in involution if all pairs of rays conjugate with respect to k^2 are correlated to each other

(Art. 207). If the involution thus constituted is rectangular its centre has a particular significance for this curve, and is called a *focus* of the curve.

We may then state the following definition :

A focus of a curve of the second order is such a point of the plane that any two rays through it which are conjugate with respect to the curve are at right angles.

227. A focus cannot lie outside the curve, for then the involution whose centre it is would have two double rays, namely, the two tangents to the curve which pass through it.

Every focus F lies upon an axis of the curve, in other words, the diameter passing through F is an axis, since it is perpendicular to its conjugate chord passing through F.

The straight line joining two foci, F and F_1, is an axis of the curve, since it is conjugate to the two perpendiculars which can be erected to it in F and F', its pole thus being the infinitely distant point of intersection of these perpendiculars (Art. 160).

228. The centre of a circle is a focus. Two rays which are conjugate with respect to a circle intersect at right angles only if one or both are diameters; whence it is easily seen that a circle has no focus except the centre. The circle, consequently, will be excluded from the following investigation.

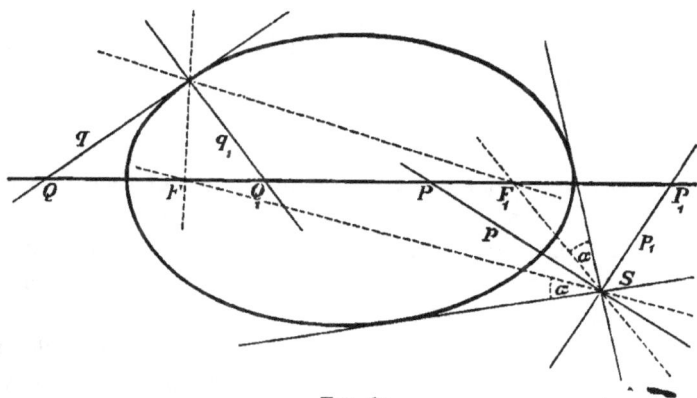

FIG. 69.

229. To any straight line p lying in the plane of a curve of the second order there is a conjugate normal straight line p_1 (Fig. 69), namely, the perpendicular which can be let fall upon it from its pole.

Let now a be an axis of a given conic and let it be cut obliquely by p and p_1 in the points P and P_1 respectively. Then each ray of the sheaf P stands at right angles to its conjugate ray of the sheaf P_1. For the sheaves P and P_1 are projectively related if to each ray of the one sheaf the conjugate ray of the other is correlated (Art. 144); but since, if A denote the infinitely distant pole of the axis a, the three rays a, PA, p, of P intersect their conjugate rays P_1A, a, p_1, of P_1 at right angles, the sheaves P and P_1 generate a circle upon PP_1 as diameter, and hence any two conjugate rays of these sheaves are at right angles to each other. From this there results the first half of the following theorem :

Corresponding to any point P *in an axis* a *of a curve of the second order there is a point* P₁ *such that conjugate rays through* P *and* P₁, *respectively, intersect at right angles. If all such pairs of points are coördinated they form an involution on the axis* a.*

The second half of the theorem arises from the following considerations :

The two sheaves of parallel rays of which the one has the direction of the ray p and the other that of the ray p_1 are related projectively to each other (Art. 144) by correlating to each ray of the one sheaf the ray of the other conjugate to it. The straight line a is cut by these sheaves in two projective ranges of points, but these are in involution since any two of its points as P and P_1 correspond to each other doubly.†

If this involution a has two double points, each of these is a focus of the curve ;‡ if a has no double points, then each of the two points from which a is projected by a rectangular involution sheaf (Art. 224) is a focus of the curve; for, every pair of conjugate rays through such points intersect at right angles.

In the latter case the foci lie in the axis of the conic different from a and form the double elements of an involution lying upon

---

* That there can be but one such involution established upon the axis a follows from the fact that any ray p has but one conjugate normal p_1.—H.

† Any ray through P is conjugate to its normal through P_1. The ray through P having the direction of p is conjugate therefore to the ray through P_1 having the direction of p_1, and also the ray through P having the direction of p_1 is conjugate to the ray through P_1 having the direction of p. The points P and P_1 therefore correspond doubly in the projective ranges of points which form sections of the parallel sheaves of rays.—H.

‡ The involution cannot have more than two double points, hence there cannot be more than two foci upon one axis.—H.

this second axis, which may be constructed in the same way as was the first.

230. No curve of the second order has more than two foci; for any straight line joining two foci is an axis of the curve (Art. 227), hence all foci must lie upon one axis or upon the other, and it was shown in the last article that not more than two foci can lie upon one axis. That a conic in general has two foci is clear from the fact that if the involution considered in Article 229 has no real double points, then two points exist on the other axis answering the conditions for foci (Art. 224).

That axis a of an ellipse or hyperbola upon which the two foci of the curve lie is called the 'principal' or 'major' axis, the other the 'conjugate' or 'minor' axis.

231. *If about a right-angled triangle whose hypotenuse lies in the minor axis of a curve of the second order, and whose other sides are polar conjugates with respect to the curve, a circle be circumscribed, it intersects the major axis in the foci of the curve.*

This theorem follows immediately from the preceding article.

232. The hyperbola is intersected by its principal axis; for upon that axis by which it is not cut the foci could not lie, since they lie within the curve. The foci of an ellipse or a hyperbola are equally distant from the centre of the curve (Art. 223); for in the involution a, whose double points the foci are, the centre is conjugate to the infinitely distant point, since the minor axis is conjugate to all straight lines normal to it. In general, the foci are harmonically separated by any two conjugate lines which are at right angles to each other.

If the curve of the second order is a parabola and a its axis, the two projective sheaves of parallel rays of which we have already spoken have the infinitely distant straight line as a self-corresponding ray, since it is self-conjugate, being tangent to the parabola. In the involution a, therefore, one double point coincides with the infinitely distant point, which, accordingly, is, looked upon as a focus (ideal) of the parabola. The parabola has therefore only one actual focus which, as second double point of the involution a, bisects the segment between any two coördinated points P and P_1.

233. Suppose now F and F_1 are the two foci of a curve of the second order, of which in the case of the parabola the one lies infinitely distant in the direction of the parallel diameters. Any

two conjugate lines SP and SP_1 which are perpendicular to each other (Fig. 69) are harmonically separated by the points F and F_1, and hence by the lines SF and SF_1; they therefore bisect the angles between SF and SF_1 (Art. 68). If S is a point of the curve one of the straight lines SP, SP_1, touches the curve; if S lies outside the curve, then SP and SP_1 bisect also the angles between the two tangents which can be drawn from S to the curve, since these likewise are harmonically separated by SP and SP_1 (Art. 143). Thus we obtain the theorems:

Any tangent to a curve of the second order forms equal angles with the two straight lines which join its point of contact to the foci of the curve. If two conics, therefore, have the same foci (are confocal) the two tangents at a real point common to both conics intersect at right angles.

If the point of intersection of two tangents to a conic be joined to its foci, the one line forms with one tangent the same angle as the other line forms with the other tangent.

234. The polar f of a focus F of a curve of the second order is called a 'directrix' of the curve. There are two directrices for

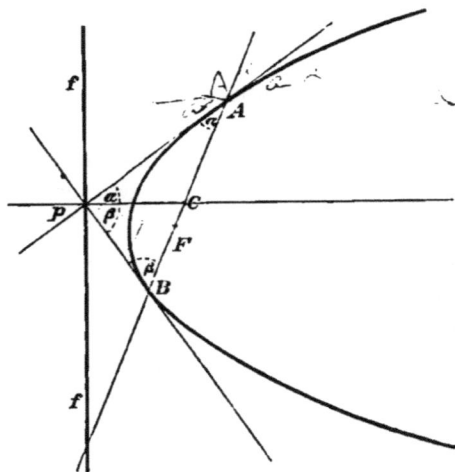

FIG. 70.

the ellipse and the hyperbola, and one actual directrix for the parabola. For the latter this theorem holds true:

Two tangents to a parabola PA *and* PB *are perpendicular to each other if their point of intersection* P *is on the directrix.*

In this case, namely, the polar of P passes through the focus F (Fig. 70) and contains the points of contact A and B of the two tangents, and each of these tangents forms the same angle with AB as with an arbitrary diameter. Consequently, in the triangle APB the sum of the angles A and B equals the sum of the angles which PA and PB make with the diameter PC passing through P, *i.e.* equal to the angle P; and since the angles A, B, and P together make up two right angles, then must P be a right angle.

235. Other properties of the focus of a curve of the second order worthy of note arise from its definition, in accordance with which any two conjugate rays through a focus intersect at right angles, among them the theorem :

" The segment of a tangent which lies between its point of contact " and a directrix is subtended at the corresponding focus by a right " angle."   -

That is to say, the straight lines bounding this angle are conjugate rays through the focus F, since the point in which the tangent is cut by the directrix f has the straight line joining F to the point of contact as its polar.

236. Let TA and TB be any two tangents to a curve of the second order (Fig. 71), so that AB is the polar of the point T; then the point of intersection P of AB and the directrix f is the pole of the straight line TF and PF is at right angles to TF since these two lines are conjugate. At the same time FA and FB are harmonically separated by FT and FP, since A and B are harmonically separated by P and FT. Consequently the supplementary angles formed by FA and FB are bisected by FT and FP (Art. 68) ; or,

" If a focus of a curve of the second order be joined to the points " of contact of two tangents to the curve, and to their point of " intersection, this latter line makes equal angles with the two former."

If through A and B (Fig. 71) straight lines be drawn parallel to FT which cut the directrix f in the points A_1 and B_1 respectively, then A_1 and B_1 are harmonically separated by P and FT. The angles between FA_1 and FB_1, therefore, are also bisected by FT and FP. From this it follows that the triangles A_1AF and B_1BF are mutually equiangular and hence similar, so that

$$FA : AA_1 = FB : BB_1.$$

The segments AA_1 and BB_1 form equal angles with the directrix *f*, and are therefore proportional to the perpendiculars which can be let fall from *A* and *B* upon *f*.

Hence, also, $FA : AA_2 = FB : BB_2$.

Since *A* and *B* are two points of the curve chosen at random, we have the theorem:

The distances of an arbitrary point of a curve of the second order from a focus and from the corresponding directrix have a constant ratio to each other (Pappus).

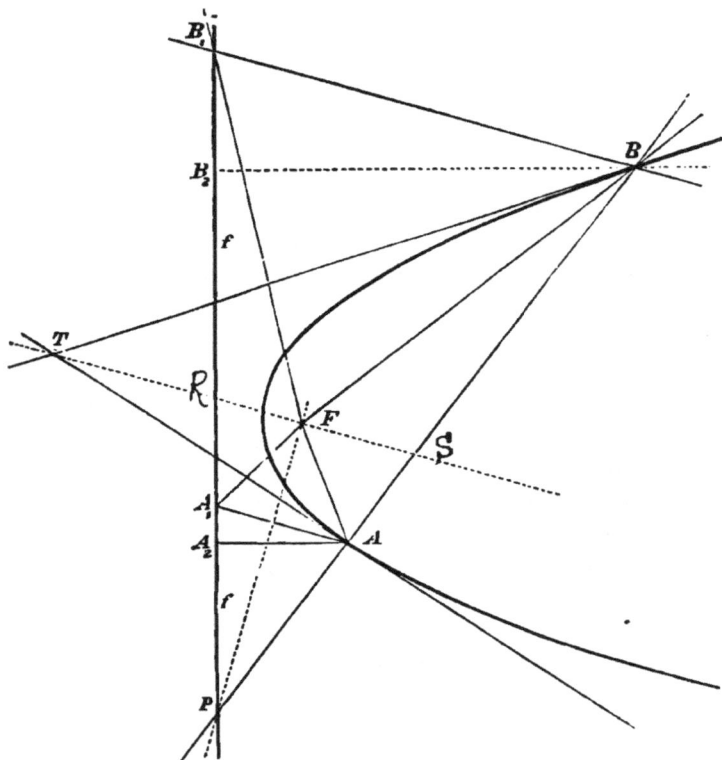

FIG. 71.

In the parabola the value of this ratio is unity, *i.e.*, the two distances are equal; for the vertex is just as far distant from the focus as from the directrix, since it is harmonically separated by them from the infinitely distant point of the parabola. By making use of the vertex it is easily shown that this ratio is less than unity in the ellipse and greater than unity in the hyperbola; and since a curve of the second order is divided into two symmetrical parts by either of its axes, the ratio has the same value for the one focus and its directrix as for the other. If then r and r_1 are the distances of any point of the curve A from the two foci F and F_1 (Figs. 72 and 73), d and d_1 its distances from the two corresponding directrices f and f_1, then

$$\frac{r}{d} = \frac{r_1}{d_1} = \text{constant},$$

wherever the point A may lie.

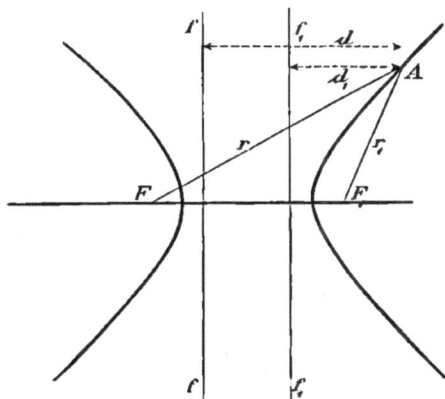 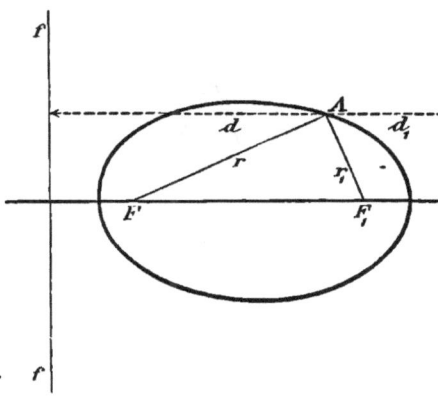

FIG. 72.   FIG. 73.

Also the quantity $\dfrac{r \pm r_1}{d \pm d_1}$ equals this same constant ratio. But in the ellipse $d + d_1$, and in the hyperbola $d - d_1$, is constant, namely, it is the distance between the two directrices. Therefore the sum $r + r_1$ must be constant in the ellipse and the difference $r - r_1$ constant in the hyperbola, that is to say:

The sum of the distances of any point of an ellipse from the foci is constant.

The difference of the distances of any point of an hyperbola from the foci is constant.

It may be found without difficulty that this constant sum or difference is equal to the segment $2a$ between the extremities of the major axis, and that the ellipse encloses a greater segment of its major axis than of its minor axis.

237. If two points are symmetrically situated with respect to a straight line, *i.e.* if the straight line joining them is at right angles to the given line and is bisected by it, then either of these is called the 'inverse' of the other with respect to the straight line. The following theorem, then, is true for the ellipse and the hyperbola:

"The points inverse to a focus F_1 with respect to the several "tangents to an ellipse or hyperbola lie upon a circle of radius $2a$ "whose centre is the other focus F."

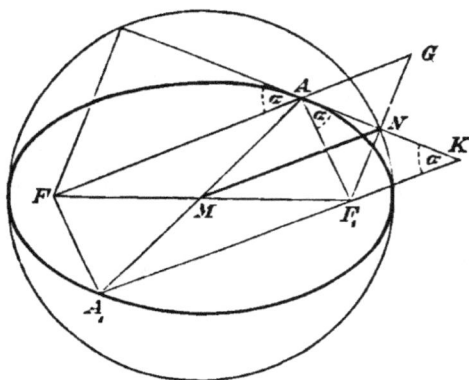

FIG. 74.

If, namely, G is the inverse of F_1 with respect to an arbitrary tangent whose point of contact is A (Fig. 74), then F, A, and G lie in one straight line, since AF_1, and consequently also AG, forms with the tangent the same angle as does AF (Art. 233). And moreover, since the points F_1 and G are equally distant from A, then is FG equal to the sum (or difference) of FA and F_1A, i.e. equal to the constant $2a$ as was shown above.

If N be the foot of the perpendicular let fall from F_1 upon the tangent, it is the middle point of F_1G, and the centre M of the curve bisects the segment F_1F; consequently, MN is parallel to FG and equals $\frac{1}{2}FG$ or a; that is to say,

"The points of intersection of all tangents to an ellipse or "hyperbola with the perpendiculars let fall upon them from a focus, "lie upon a circle which has the major axis as diameter."

238. If the parabola is regarded as the limiting case of an ellipse or hyperbola, for example, as an ellipse one of whose foci lies infinitely distant, we obtain the following theorem:

"The points of intersection of all tangents to a parabola with

"the perpendiculars let fall upon them from the focus F lie upon
"the tangent at the vertex of the parabola."

In order to prove this draw through N, the point of intersection
of an arbitrary tangent with the tangent at the vertex (Fig. 75),
a line NF_1 parallel·to the axis, and draw the straight line NF to
the focus; then the angles FNA and SNF_1 are equal (Art. 233),
since NF_1 passes through the second (infinitely distant) focus;
and since SNF_1 is a right angle, FNA must be also.

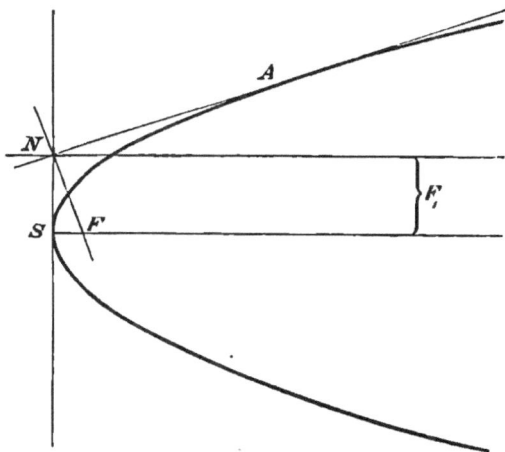

FIG. 75.

This theorem gives us a very simple means of constructing the
focus of a parabola.   A simple construction for the foci of an
ellipse or hyperbola is the following:

Draw the tangents at the extremities of the major axis; these
intersect any third tangent in two points P and Q, which are
projected from any point of the major axis by two conjugate
rays (Art. 146).   In order, then, to obtain the foci, at either of
which these two conjugate rays are perpendicular to each other,
we describe a circle upon PQ as diameter; the points of intersection of this circle with the major axis are the required foci
(Apollonius).

239. If two tangents TA and TB of a curve of the second
order are cut by a third tangent in the points A_1 and B_1, respectively, A, B, and C being the three points of contact, and F an
actual focus of the curve (Fig. 76), then for the angles subtended

at F by segments of the tangents the following equations are true :

$$\angle B_1FC = \angle BFB_1 = \tfrac{1}{2}\angle BFC,$$
$$\angle CFA_1 = \angle A_1FA = \tfrac{1}{2}\angle CFA ;$$

consequently, $\angle B_1FC + \angle CFA_1 = \tfrac{1}{2}(\angle BFC + \angle CFA),$

or $\angle B_1FA_1 = \angle BFT = \angle TFA.$

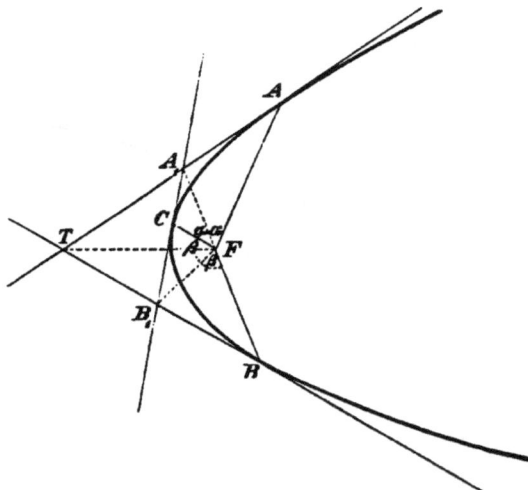

FIG. 76.

If, then, the first two tangents TA and TB remain fixed, the angle B_1FA_1, subtended at F by the segment of the third tangent lying between the first two, has a constant magnitude however this third tangent may be chosen. If the third tangent glides along the curve, A_1 and B_1 describe two projective ranges of points upon the fixed tangents, while the rays FA_1 and FB_1 describe projective sheaves of rays about F; therefore,

"The projective ranges of points in which any two tangents of "a curve of the second order are cut by the remaining tangents are "projected from either focus of the curve by two equal and directly "projective sheaves of rays."

This theorem holds true also if the curve be a parabola or a circle, and one focus be the infinitely distant point of the former or both coincide in the centre of the latter. If, then, three tangents and a focus of a curve of the second order be given; any required number of other tangents can be easily constructed.

If the given curve is a parabola the moving tangent A_1B_1 may

M

become infinitely distant, in which case, the straight lines FA_1 and FB_1 form the same angle with each other as do the fixed tangents TA and TB. The quadrangle B_1TA_1F may therefore be inscribed in a circle, and hence,

"The circle which is circumscribed to any tangent triangle of a "parabola passes through the focus."

If, then, we circumscribe circles about the four triangles which are formed of the sides of a quadrilateral, these have one point in common, namely, the focus of the parabola inscribed to the quadrilateral.

If the curve of the second order is a hyperbola and the two fixed tangents TA and TB are its asymptotes, then A and B are infinitely distant points, hence TB and FB are parallel and the angle B_1FA_1 ($= BFT$) is equal to one of the two angles which the asymptote TB makes with the major axis FT; the other is equal to the angle $A_1F_1B_1$ which A_1B_1 subtends at the second focus F_1. Consequently, B_1FA_1 and $A_1F_1B_1$ are supplementary angles and $B_1FA_1F_1$ may be inscribed in a circle. Hence:

"The two foci of a hyperbola lie upon a circle with the points "in which an arbitrary tangent is cut by the two asymptotes. The "centre of this circle and the centre of the hyperbola lie upon a "second circle with these same two intersections."

EXAMPLES.

1. To construct the focus of a parabola :

(a) The focus lies upon that perpendicular to any tangent which can be drawn from its point of intersection with the tangent at the vertex (Art. 238).

(b) The focus bisects that segment of the axis which lies between any two conjugate lines at right angles to each other (Art. 233).

(c) Any tangent to a parabola forms the same angle with a diameter as with the line joining its point of contact to the focus (Art. 233).

(d) Any circle which is circumscribed about a triangle whose sides are tangent to a parabola passes through the focus (Art. 239).

2. To construct the directrix of a parabola :

(a) The directrix is the polar of the focus.

(*b*) Pairs of tangents at right angles to each other intersect on the directrix (Art. 234).

(*c*) The orthocentres of all circumscribed triangles lie upon the directrix (Example 12, p. 162).

3. The directrices of all parabolas which can be inscribed in a triangle pass through the orthocentre of that triangle, and their foci lie upon the circumscribed circle. If perpendiculars be dropped from any point of this circle to the sides of the triangle, their points of intersection with the sides lie in a straight line, viz. the tangent at the vertex of one of the inscribed parabolas.

4. To construct the two foci of an ellipse or hyperbola :

(*a*) As double points of an involution lying upon the major axis (Art. 229).

(*b*) By means of the tangents at the extremities of the major axis. These determine upon any third tangent a segment which subtends a right angle at either focus (Art. 238).

(*c*) By means of the circle which touches the curve at the extremities of the major axis. If perpendiculars be drawn to any tangent at the points in which it intersects this circle, they will pass through the two foci (Art. 237).

(*d*) By constructing a right-angled triangle whose hypotenuse lies in the minor axis and whose sides are conjugate. The circle circumscribing such a triangle intersects the major axis in the two foci (Art. 231).

(*e*) With the aid of the theorem that the sum (or difference) of the focal distances of a point of the curve is constant (Art. 236).

5. Construct a curve of the second order having given one focus, the corresponding directrix, and one point or one tangent.

6. Construct a curve of the second order having given one focus and (1) three tangents, or (2) two tangents and the point of contact in one of them (Art. 239). In either case the second focus can be immediately determined.

7. If in a plane the foci of any curve of the second order inscribed to the triangle *ABC* are correlated to each other, an involution of the second order is established among the points of the plane, of which *A*, *B*, and *C* are the three singular points ; that is, if the one focus describes a straight line *u* the other will describe a conic passing through *A*, *B*, and *C*, and projective to *u* (Art. 233). The straight lines bisecting the angles of the triangle are coördinated to themselves. Every straight line of the plane is an axis of one of the inscribed curves.

8. Construct a curve of the second order having given the two foci and one point or one tangent.

9. The angles made by the two tangents from a point P to the several curves of the second order of a confocal system are all bisected by two fixed straight lines at right angles, namely, by the tangents to the two curves of the system which pass through P.

10. The poles of a straight line g with respect to a system of confocal conics lie in a straight line perpendicular to g. This line is harmonically separated from g by the two foci.

11. Each pair of opposite sides of a quadrangle determined by the three vertices of a self-polar triangle of a curve of the second order and its orthocentre, are harmonically separated by the foci of the curve.

12. Given two points, A and B, and one focus F of a curve of the second order, the other focus F_1 lies upon one of the two conics passing through F, of which A and B are the foci. For, since $AF \pm BF$ is given, $AF_1 \pm BF_1$ must be of constant magnitude.

The directrix corresponding to F intersects the straight line AB in one or other of the two points through which pass the lines bisecting the angles formed by FA and FB.

13. The product of the distances of a focus from two parallel tangents to an ellipse or hyperbola is constant (Art. 237); so also is the product of the distances of the two foci from one tangent.

14. Construct an ellipse with axes of given length (a) after the method proposed in Ex. 15, p. 125; (b) by making use of the tangents at the four vertices; (c) with the aid of the foci, which may be determined when the lengths of the axes are given.

15. Construct a parabola having given the focus and the directrix, or the focus, the axis, and one point or one tangent.

16. The centres of all circles which touch two given circles lie upon two confocal curves of the second order (compare Art. 236). The centres of the two given circles are the foci of these two curves.

17. All points of a plane which have equal distances from a straight line and a circle lie upon either one or the other of two parabolas (compare Ex. 16).

18. If one side of an angle of given magnitude constantly touches a curve of the second order while the other rotates about one of its foci, the vertex of the angle describes a circle which touches the curve at two points (Art. 237); if, however, the given curve is a parabola the vertex describes a straight line tangent to the curve.

LECTURE XIV.

PROBLEMS OF THE SECOND ORDER. IMAGINARY ELEMENTS.

240. Our investigations have often brought us to problems which admit in general of two solutions, and which cannot be solved by the exclusive application of linear constructions, but require the use of a form of the second order. To this class belong among others the problems "to determine the self-corresponding points in two projective ranges which are superposed," and "to determine the double elements of a form in involution."

All such problems may be reduced to the following:

Two ranges of points k^2 and k^2_1 of the second order lie upon the same curve and are projectively related; it is required to determine their self-corresponding points.

Let A, B, C (Fig. 77), be any three points of k^2, and A_1, B_1, C_1, the corresponding points of k^2_1. If we project the range k^2_1 from A and the range k^2 from A_1 we obtain two projective sheaves $A(A_1B_1C_1)$ and $A_1(ABC)$ which have the ray AA_1 self-corresponding. The points of intersection of pairs of homologous rays of these sheaves lie consequently upon a straight line u, namely, upon that

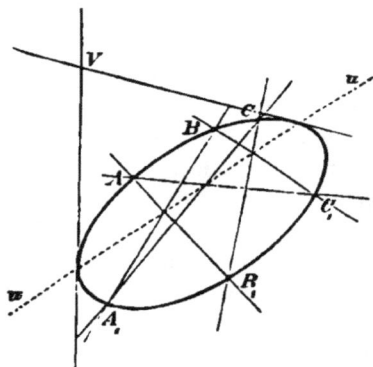

FIG. 77.

line which joins the intersection of AB_1 and A_1B with the intersection of AC_1 and A_1C. The points which u has in common with the curve of the second order are the required points, for in the ranges k^2 and k^2_1 these points are self-corresponding. According, then, as u cuts, touches, or does not meet the curve do we obtain two, one, or no such self-corresponding points.

In accordance with Pascal's theorem the intersection of the straight lines BC_1 and B_1C also lies upon the straight line u; for upon this line the three pairs of opposite sides of the hexagon $AB_1CA_1BC_1$, which is inscribed in the conic, intersect. We obtain this same straight line, then, by projecting the ranges k^2_1 and k^2 from the points B and B_1, or from C and C_1, respectively.

If in general P and Q are any two points of k^2 and P_1 and Q_1 the corresponding points of k^2_1, then the points of intersection of PQ_1 and P_1Q lie upon the straight line u. If then three points of k^2 together with the corresponding points of k^2_1 are given, the straight line u can be constructed without difficulty.

241. If the ranges k^2 and k^2_1 are in involution, u is the axis of involution, and the curve is cut by u in the two double points, if such there are. In this case we need to know only two pairs of coördinated points A, A_1, and B, B_1, in order to construct u; for, to the points A, B, A_1, of k^2 correspond the points A_1, B_1, A, respectively, of k^2_1, and u passes through the two points in which the straight lines AB_1 and AB are cut by the straight lines A_1B and A_1B_1, respectively (Figs. 60 and 61).

The pole of u, in which AA_1 and BB_1 intersect, is the centre of involution, and if from this point two tangents can be drawn to the curve, these touch the curve in the double points of the involution.

242. The different cases of the following general problem may easily be reduced to the preceding :

"To determine the self-corresponding elements of two projective "elementary forms which are superposed."

If, for example, the elementary forms are two cones or two reguli of the second order, they may be cut by an arbitrary plane in two projective ranges of points which lie upon the same curve of the second order; if they are two sheaves of rays of the second order, then the curves of the second order which are enveloped by them are also superposed and are projective, so that we need only determine the points which the curves have self-corresponding in order to obtain immediately the two self-corresponding rays, the tangents at these points.

If two projective sheaves of rays of the first order are concentric and lie in one plane, and are intersected by a curve of the second order passing through their common centre, in two projective ranges of points k^2 and k^2_1, the two rays which are self-corresponding in the sheaves pass through the two points which k^2 and k^2_1 have self-corresponding.

If the two elementary forms to be considered are ranges of points v and v_1 (Fig. 78) which lie upon the same straight line, the case can be reduced to the preceding one by projecting the ranges from any point S. Pass, then, through S a conic (for example, a circle), in the plane Sv, and project any three points

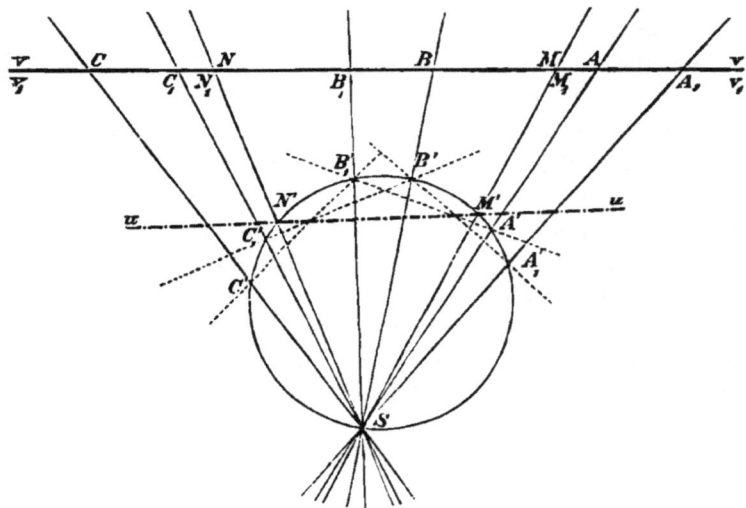

FIG. 78.

A, B, C, of v and their corresponding points A_1, B_1, C_1, of v_1 from the point S upon this curve. By this means we obtain the curve points A', B', C', and A'_1, B'_1, C'_1, and can determine immediately the straight line u upon which the points of intersection of $A'B'_1$ and A'_1B', of $B'C'_1$ and B'_1C', and of $C'A'_1$ and C'_1A' lie. If, then, the conic is cut by u in two points M' and N', and we project these from the point S upon the straight line v, we obtain the two points M and N of v which coincide with their corresponding points M_1 and N_1 of v_1.

If the conic has only one point or no point common with u, there is only one or no self-corresponding point of the ranges v and v_1.

For the sheaf and the cone of the second order the general problem might be solved directly, without the use of the curve of the second order; the general problem stated above can therefore be solved for the case of any of the one-dimensional primitive forms with the help of any one of the elementary forms of the second order. The most convenient solution, however, is the one just given, since curves of the second order (circles in particular) may easily be constructed.

243. As is well known, many of the advances in mathematics are intimately connected with the effort to remove exceptional cases from general theorems and principles, and with the attempt to reconcile different theorems to the same point of view, by immediate extension or by the introduction of new concepts.

Thus arithmetic was essentially enriched by the introduction of negative, of irrational, and, finally, of imaginary numbers; without the latter the important theorem, "an equation of the nth order has n roots," with its numerous applications (for example, in analytical geometry) would be wholly false. In the same way the introduction of infinitely distant elements into modern geometry has proved fruitful in the highest degree.

Problems of the second order have given the first occasion for the introduction into synthetic geometry of 'imaginary' points, lines, and planes; and, to have founded the purely geometric theory of imaginary elements and to have brought it to a high degree of completeness is undoubtedly one of the greatest services which Von Staudt has rendered. From the nature of the subject this theory must, in synthetic as well as in analytical geometry, give up all claim upon the powers of intuition; consequently, I shall confine myself at this point to the presentation of only the first principles of the theory of geometric imaginaries.

244. We shall define imaginary elements by the following proposition which, at the same time, summarizes the results of earlier investigations:

Two projective elementary forms which are superposed but are not identical have two 'real' or two 'conjugate imaginary' self-corresponding elements. If the self-corresponding elements are real they may coincide.

We say that the self-corresponding elements are 'imaginary' whenever they do not really appear. In all my previous lectures only 'real' elements have been considered. An involution always has from this point of view two (real or imaginary) double elements. Moreover, we can say that—

A curve of the second order has two points in common with *every* real straight line of its plane;	Two tangents to a curve of the second order pass through *every* real point of its plane;

and only when the different cases comprised under these theorems are to be distinguished from each other do we add—

These two points are imaginary, or real, or they coincide, according as the straight line lies wholly outside, or cuts, or touches the curve.	These two tangents are imaginary, or real, or they coincide, according as the point lies within, or without, or upon the curve.

If the curve and the straight line are completely given we shall always regard their common points as being determined. But if, for example, only five points A, B, C, D, E, of the curve are given, we imagine it to be generated by two projective sheaves of rays $A(CDE)$ and $B(CDE)$; these are intersected by the straight line in two projective ranges of points which have the required points of intersection with the curve as self-corresponding points. The two points can be determined by the method given above (Art. 242) if any one curve of the second order is completely known. We can also ascertain by this means to which variety a conic determined by five points belongs; for, a curve of the second order is a hyperbola, ellipse, or parabola according as the two points which it has in common with the infinitely distant straight line are real and different, imaginary, or coincident.

The following problem of the second order may be solved by the same method:

Of a curve of the second order there are given—

Four points and the tangent at one of them, or three points and the tangents at two of them; it is required to determine the two points which the curve has in common with any given real straight line of its plane.	Four tangents and the point of contact on one of them, or three tangents and the points of contact on two of them; it is required to determine the two tangents which pass through any given real point of its plane.

If it is required to determine the common points of a straight line and a curve of the second order of which five tangents are given, we find first the points of contact in these tangents and so reduce this problem to the one just solved.

245. Von Staudt distinguished between the two conjugate imaginary double elements of an elliptic involution AA_1, BB_1, . . . by connecting with the form a determining sense ABA_1 or A_1BA.

Without going into the matter more minutely we can enunciate the following theorems and definitions upon the basis of what has already been said :

An imaginary point always lies upon a real straight line ; this line also contains the conjugate imaginary point.

An imaginary plane always passes through a real straight line ; through this straight line passes also the conjugate imaginary plane.

An imaginary straight line 'of the first kind' always passes through a real point and lies in a real plane with its conjugate imaginary line ; namely, the point and the plane are the bases of the projective sheaves of the first order which have the two conjugate imaginary straight lines self-corresponding.

Two projective reguli of the second order which are superposed have two real straight lines, which may however coincide, or two conjugate imaginary straight lines 'of the second kind,' self-corresponding. These imaginary lines of the second kind are distinguished from the imaginary lines of the first kind in that they can be cut by no real plane in real points and can be projected from no real point by real planes. A real plane, namely, cuts the two projective reguli in two projective ranges of points of the first or second order which lie upon the same base and have real self-corresponding points only if the reguli have real self-corresponding lines (passing through them). Thus there exists only one kind of imaginary points or planes, but, on the other hand, two kinds of imaginary straight lines.

246. Whenever we make mention simply of points, straight lines, and planes, we shall refer as heretofore to real elements, unless the contrary is expressly stated or is evident from the context. This applies in particular to the following problem of the second order :

" In a plane there are given two simple polygons ; it is required " to construct a third polygon which is inscribed to one of these " and circumscribed about the other."

Or to speak more definitely,

" To construct an n-point whose vertices lie in order upon n

"given straight lines u_1, u_2, u_3, ... u_n, and whose sides pass in order "through n given points S_1, S_2, S_3, ... S_n, of a plane."

Project the range of points u_1 from the point S_1 upon the straight line u_2; next, project the range u_2, *i.e.* the projection of u_1, from S_2 upon the straight line u_3, then the range u_3 from the point S_3 upon the straight line u_4, and so on, till finally we project the range u_n from the point S_n upon the straight line u_1. By this means · we obtain $n + 1$ projective ranges of points of which each is a projection of the preceding, and of which the first and last lie upon one and the same straight line u_1. Each point which the first and last ranges have self-corresponding can be chosen as the first vertex of the required n-point, and a solution of the problem immediately presents itself.

In general there are then at most two n-points which satisfy the conditions of the problem.

If in particular cases the two projective ranges lying in u_1 have more than two and consequently all their points self-corresponding, then there is an infinite number of solutions of the problem.

The condition that the straight lines u_1, u_2, u_3, ... u_n, shall lie in one and the same plane with the points S_1, S_2, S_3, ... S_n, is moreover not necessary; it is sufficient if S_1 lies in a plane with u_1 and u_2, S_2 in a plane with u_2 and u_3, and so on, till finally S_n lies in a plane with u_n and u_1. The two resulting n-points would then be gauche.

247. In this connection belongs also the problem, "To find a "straight line which intersects four given straight lines a, b, c, d, "when no two of them lie in one plane."

Relate the sheaves of planes a and b perspectively to the range of points c, and let d intersect them in two projective ranges of points. Through each of the points which these ranges have self-corresponding a straight line passes which is cut also by a, b, and c, since it lies in two homologous planes of the sheaves a and b. If the straight lines a, b, c, d, belong to one and the same regulus, then the problem has an infinite number of solutions; in general, however, there are but two. This problem might be stated thus:

"A ruled surface of the second order is given by three straight "lines a, b, c, of one of its reguli; it is required to find the points "which it has in common with any fourth line d."

248. One of the most important problems of the second order is the following:

"Two involutions lie upon the same base; it is required to "determine two elements which are coördinated to each other in "both involutions."

Suppose the two involutions lie on the same curve of the second order, and, in the one involution, to some two points α and β of the curve the points α_1 and β_1 respectively are coördinated, while in the other involution, to the points A and B are coördinated the points A_1 and B_1, respectively; then we seek the two centres of involution U and V, so that all pairs of conjugate points of the one involution lie in a straight line with U, and all pairs of such points of the other lie in a straight line with V. If, then, the straight line UV intersects the curve in two points X and X_1, these points will be coördinated to each other in both involutions. If UV touches the curve, the point of contact is a double point of each involution. Finally, if UV lies wholly outside the curve, there is no real point which is coördinated either with itself or with a single other point in both involutions.

This latter case can happen, however, only when each of the two involutions has two real double points, *i.e.* both are hyperbolic, since only then do both points U and V lie outside the curve; and further, when the polars of U and V intersect within the curve, namely, in the pole of UV, *i.e.* when the double points of the one involution are separated by those of the other.

If the problem should relate to two concentric sheaves of rays in involution we may intersect them with a curve of the second order which passes through the common centre, and in a similar way can reduce any chosen case of the general problem to that just now treated.

The result last obtained holds good, therefore, not only for two point involutions which lie upon the same curve of the second order, but it can be stated for a general case thus :

"If two involutions are superposed there will always be two "elements which are coördinated to each other in the one involution

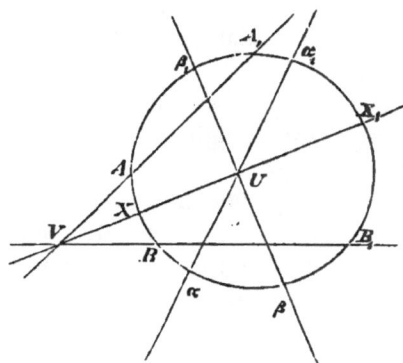

FIG. 79.

"as well as in the other; these elements are (conjugate) imaginary
"only in case both involutions are hyperbolic and the double
"elements of the one are separated by the double elements of the
"other. If the two doubly conjugate elements coincide, the
"involutions have one double element in common."

Suppose, for example, the involutions are two sheaves of rays
of the first order and one of them is rectangular (Art. 224), then
this theorem follows as a special case:

"In any sheaf of rays of the first order in involution there are
"always two real conjugate rays which are at right angles; these
"are called the 'axes' of the involution sheaf."

249. We have thus proved in an entirely different manner the
theorem that an ellipse or hyperbola has two conjugate diameters
at right angles, *i.e.* two axes (compare Art. 159); for, its diameters
form an involution if pairs of conjugate diameters are coördinated
to each other.

This result naturally belongs to metric geometry, as does also
the allied problem:

"To construct the axes of a curve of the second order of which
"two pairs of conjugate diameters are given."

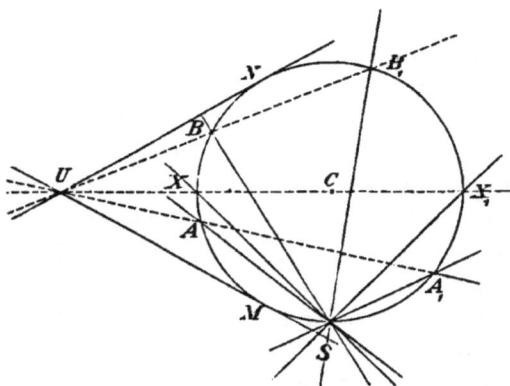

FIG. 80.

Through the centre S of the curve (Fig. 80) in which the given
diameters intersect, pass an arbitrary circle; let this cut the one
pair of conjugate diameters in the points A and A_1, the other
pair in the points B and B_1. Then the points of the circle are
paired in involution by means of the sheaf of diameters, and

the point U in which AA_1 and BB_1 intersect is the centre of involution with which each pair of coördinated points of the circle lie in a straight line. If then we pass a straight line through U and the centre C of the circle, this will cut the circle in two coördinated points X and X_1 which are projected from S by the required axes SX and SX_1.

If U lies outside the circle there exist two real double points of the involution M and N; these are projected from U by two tangents to the circle, but from S by the asymptotes of the hyperbóla to which the given conjugate diameters belong (Art. 161). How could the axes be constructed if instead of the circle an arbitrary curve of the second order passing through S were given?

250. In what follows we shall many times need to solve the problem :

" In a sheaf of rays in involution to determine two conjugate "rays which are harmonically separated by two given points in the "plane."

In order that this problem may not be impossible we assume that the given points M and N do not lie in a straight line with the centre of the sheaf, and that through neither of them passes a double ray of the sheaf. If then we project the involution range of points, of which M and N are the double points, from S by a second involution sheaf, we shall have only to find those two rays which are coördinated to each other in both sheaves, and these will be the rays required.

This problem may be carried over to other elementary forms and stated in different ways. Instead of the sheaf S, for example, might be given an involution range upon the straight line MN; if in this case one of the given points, say N, lies infinitely distant the problem becomes :

" In a point involution of the first order it is required to determine " two conjugate points which have equal distances from a given " point M of the straight line."

251. A triangle ABC and an involution sheaf of rays of the first order F, of which no double ray passes through a vertex of the triangle, are given in a plane. It is required so to circumscribe a curve of the second order about the tri-

A triangle and a point involution of the first order, of which no double point lies upon a side of the triangle, are given in a plane. It is required so to inscribe a curve of the second order to the triangle that all pairs of corresponding

angle that all pairs of corresponding points of the involution are con-
rays of the involution are con- jugate with respect to the curve.
jugate with respect to the curve.

In order that the problem may be possible it is necessary that the centre of the involution F should coincide with no vertex of the triangle ABC (Art. 142); we assume therefore that at least two of the sides of the triangle, say AB and AC, do not pass through F (Fig. 81). Upon each of these sides AB and AC there

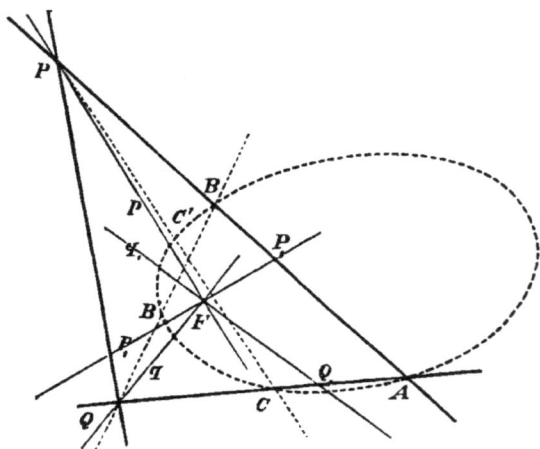

FIG. 81.

must lie one point whose polar with respect to the required curve, in case the latter exists, passes through F; and since a point is harmonically separated from its polar by the curve, while, on the other hand, all pairs of corresponding rays of F are by assumption conjugate with respect to the required curve, these points may be determined as follows:

Determine in the involution F two corresponding rays p and p_1, which are harmonically separated by the points A and B, and two others q and q_1 harmonically separated by A and C (Art. 250). If these rays are imaginary, then no real curve of the second order exists which fulfils the given conditions, but this case can only happen if the involution F has two real double rays which are separated by A and B or by A and C (Art. 248). Let the straight line AB be cut by the rays p and p_1 in the points P and P_1, respectively, and let AC be cut by q and q_1 in the points Q

and Q_1, respectively. Then we must make one or other of the following suppositions:

(1) P and Q are the poles of p_1 and q_1, respectively;
(2) P and Q_1 are the poles of p_1 and q, respectively;
(3) P_1 and Q are the poles of p and q_1, respectively;
(4) P_1 and Q_1 are the poles of p and q, respectively.

Each of these four suppositions yields a solution of the given problem. If, for example, the first supposition is made, we should seek upon the straight line PC the point C' which is harmonically separated from C by P and its polar p_1, and similarly upon QB the point B' which is harmonically separated from B by Q and its polar q_1. That curve of the second order which passes through the five points $ABCB'C'$, and of which any number of points can be determined by previously given methods, fulfils, then, all the given conditions. For, it is circumscribed to the triangle ABC, and since two pairs of points of the curve, A, B, and C, C_1, are harmonically separated by P and p_1, then is P the pole of p_1 and the ray FP or p is conjugate to p_1, and likewise the ray q is conjugate to q_1, so that all pairs of corresponding rays of the involution F are conjugate with respect to the curve.

252. If the involution F has two real double rays, these of necessity being tangent to the required curve, we can state the problem thus:

About a given triangle to circumscribe a curve of the second order which shall touch two given lines in the plane.	In a given triangle to inscribe a curve of the second order which shall pass through two given points in the plane.

The above construction shows that each of these reciprocal problems has four and only four real solutions if, in the one case, the two given straight lines are separated by no two vertices of the triangle, and in the other case, if the two given points are separated by no two sides of the triangle; otherwise, there is no real solution.

253. If the double rays of the involution F are imaginary the problem has four solutions. This case happens, among other ways, if the involution is rectangular and consequently F a focus of the required curve (as was supposed in Fig. 81); incidentally, then, we have solved the problem:

"To determine the four curves of the second order which circum-
"scribe a given triangle and have a given point as focus."

254. In conclusion, a problem may here be inserted which is
clearly not of the second order, but which is closely related to
the one last discussed, namely,

In a plane are given a triangle ABC and an involution point-range of the first order u, so situated that neither double point of u lies on a side of the triangle and u passes through no vertex of the triangle. It is required so to circumscribe a conic about the triangle that all pairs of corresponding points of u are conjugate with respect to the curve.

In a plane are given a triangle and an involution sheaf of rays of the first order S, so situated that neither double ray of S passes through a vertex of the triangle and the centre S lies upon no side of the triangle. It is required so to inscribe a conic in the triangle that all pairs of corresponding rays of S are conjugate with respect to the curve.

Let the points K and M (Fig. 82), in which u is cut by the
lines AB and BC respectively, be coördinated in the involution

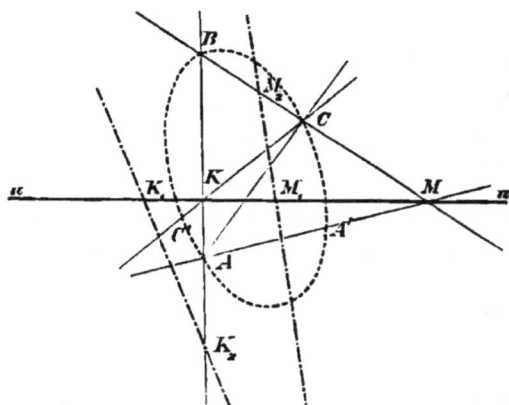

FIG. 82.

u to the points K_1 and M_1, and let K_2 be that point of AB
which is harmonically separated from K by the points A and B
of the curve, and M_2 that point of BC which is harmonically
separated from M by B and C. With respect to the required
conic, K_1K_2 is the polar of the point K, since both K_1 and K_2
are conjugate to K; similarly, M_1M_2 is the polar of M. If then
C' is the point which is harmonically separated from C by K

N

and K_1K_2, and A' the point which is harmonically separated from A by M and M_1M_2, the required curve passes through, and is determined by, the five points A, B, C, A', C'. Evidently the points K and K_1 (and similarly, M and M_1) are conjugate with respect to this curve as was required, since K is the pole of K_1K_2, being harmonically separated from the straight line K_1K_2 both by the points A, B, and C, C'.

This problem might also be stated as follows (compare Art. 244):

Through five points of a plane of which three are real and the other two either are real or are conjugate imaginaries, and of which no three lie in one straight line, to describe a curve of the second order.

To a plane pentagon, three of whose sides are real and the other two either are real or are conjugate imaginary lines, to inscribe a curve of the second order.

Each of the reciprocal parts of this problem has one solution.

EXAMPLES.

1. In a given curve of the second order inscribe a simple n-point whose sides in order pass through n given points of the plane, none of which lie upon the given curve.

2. About a given curve of the second order circumscribe a simple n-point whose vertices in order lie upon n given straight lines of the plane, none of which touch the curve.

3. In a plane there is given a simple pentagon; it is required to draw a second pentagon which is both inscribed and circumscribed to the given one.

4. Through a given point it is required to draw two straight lines which will intercept equal segments upon two given lines.

5. Choose a point in each of two given straight lines so that the line joining them subtends a right angle at either of two given points.

6. Upon a given straight line determine a segment which subtends a given angle at either of two given points.

7. Through a given point draw a straight line which bisects a given triangle, or which forms with two given straight lines a triangle of given area.

8. Circumscribe a triangle about a given triangle so that two of its vertices shall lie upon two given straight lines and the angle at the third vertex shall be of given magnitude.

9. In two projective sheaves of rays find a pair of rays at right angles in the one, which correspond, each to each, to a pair at right angles in the other (Ex. 5, p. 67).

10. In two projective ranges of points upon the same straight line find a pair of homologous points which are a given distance apart; or, in two concentric sheaves of rays projectively related find two homologous rays making a given angle with each other.

11. In two projective ranges of points of the first order find a pair of homologous segments of given lengths.

12. A curve of the second order has in general one pair of conjugate diameters which are parallel to two conjugate diameters of another curve of the second order lying in the same plane. Two hyperbolas, the directions of whose asymptotes separate each other, make an exception to this statement (Art. 248).

13. About a given quadrangle circumscribe a curve of the second order which shall touch a given straight line (Art. 221); or, to a given quadrilateral inscribe a curve of the second order which shall pass through a given point.

14. In the plane of two curves of the second order there is at least one real point U which has the same polar u with respect to both curves (Ex. 5, p. 110). The two real or conjugate-imaginary points of u which are conjugate with respect to both curves (see Art. 248) form with U a common self-polar triangle of the two curves. Two curves of the second order lying in the same plane, therefore, have in general one common self-polar triangle; this has at least one real vertex and one real side.

LECTURE XV.

255. The polar theory of cones of the second order follows by projection, as has already been said (Art. 148), from that of curves of the second order, but it is further concerned with certain theorems of metric geometry which can be developed out of the polar theory, though they can in no way be transferred from the curve to the cone by methods of projection and section; they are in part of different nature for the cone, and for it must be developed independently. Our previous investigations, however, will frequently serve us as models and furnish suggestions for further developments.

256. If a cone of the second order is chosen in a bundle of rays, all planes which pass through a ray e of the bundle are conjugate with respect to the cone to a certain plane ϵ also belonging to the bundle, where e is the pole-ray of ϵ. The ray e and that ray of the bundle which is normal to ϵ determine a plane ϵ' which is normal and at the same time conjugate to ϵ. In general there is only one such 'conjugate normal plane' for each plane ϵ of the bundle; but if a plane a is at right angles to its pole-ray a all its conjugate planes a' are normal to it. In this case a and a bisect the angles which are formed by any two rays of the cone lying in a plane with a; for a and a are normal to each other and harmonically separate the two rays (Arts. 68 and 148). The rays of the cone are thus symmetrical, two by two, with respect to the plane a, and we may call a a *plane of symmetry* for the cone; its pole-ray is usually designated a *principal axis* of the cone. A plane of symmetry for the cone thus stands perpendicular to the corresponding principal axis, *i.e.* perpendicular to its pole-ray.

257. If a plane ϵ rotates about a ray s of the bundle, its pole-ray e describes a sheaf of rays in the polar plane of s while its normal ray describes a sheaf in the plane normal to s. The two sheaves of rays thus described are projectively related to the sheaf of planes s and consequently also to each other, and generate in general a sheaf of planes of the second order. Hence:

"If, of two conjugate normal planes of the bundle, the one ϵ "rotates about an axis s, the other ϵ' will describe in general a sheaf "of the second order which contains the polar plane of s with "respect to the given cone and the plane of the bundle normal "to s."

The plane ϵ' rotates about an axis s' only if the two projective sheaves of rays have perspective position, that is, have a self-corresponding ray a. In this case s lies in a plane of symmetry a of which a is the pole-ray (the corresponding principal axis); and since a is the conjugate normal plane of the plane sa, s' must also lie in a. That is,

"In a plane of symmetry a for a cone of the second order each "ray s of the bundle is coördinated to a ray s' so that any two "planes of the sheaves s and s' normal to each other are conjugate "with respect to the cone."

This theorem is the analogue of that by the help of which (Art. 226) we obtained the foci of a curve of the second order. It brings us to the so-called focal axes of the cone of the second order, but we must first decide the question whether planes of symmetry really exist for such a cone, and if so, how many.

258. We shall show in the first place that the cone has at least one plane of symmetry.

Let Σ be the sheaf of planes which is formed by the conjugate normals of all planes passing through a ray s of the bundle. If this is of the first order, the existence of a plane of symmetry is evident. We shall suppose therefore that Σ is of the second order. To this sheaf of planes belongs every plane of symmetry for the given cone as conjugate normal to the plane passing through s and the corresponding principal axis. Now the planes η which cut the cone enveloped by Σ in two rays are separated from the remaining planes ϕ of the bundle by the tangent planes of the cone (compare Art. 136). If then t is the straight line in which the conjugate normal of one of the planes η and a plane ϕ intersect, the conjugate normals of all planes passing through t form a sheaf

T of the first or second order which has at least two and at most four real planes in common with Σ. One of these common planes is the conjugate normal of st; but every other is normal to two different planes conjugate to it, namely, to a plane in each of the sheaves s and t, and hence is normal to its pole-ray in which these two planes intersect. And since any plane normal to its pole-ray is a plane of symmetry for the cone, there exists at least one, but in general and at most three, planes of symmetry.

259. For every cone of the second order there exists then at least one plane of symmetry α and a principal axis a corresponding to it ; but it can easily be.shown that in the principal axis a two planes of symmetry always intersect at right angles. Namely, if we coördinate, two and two, those rays of the bundle lying in the plane of symmetry α which are conjugate with respect to the cone, we obtain an involution whose axes b and c are principal axes of the cone ; for, the ray b, for example, is conjugate and at the same time normal to c and to a ; its polar plane ca is consequently normal to b, and hence is a plane of symmetry of the cone. In particular, if the involution α is rectangular each of its rays is a principal axis and every plane through a is a plane of symmetry. We are easily convinced that in this special case the cone of the second order is a cone of revolution, that is, an ordinary circular cone which has a as its axis of rotation.

We have thus shown that—

" A cone of the second order has in general three planes of " symmetry which intersect at right angles in the three principal " axes a, b, c, and thus form a rectangular self-polar three-edge of " the cone. The cone of revolution alone has not only three " but an infinite number of planes of symmetry which with one " exception pass through the axis of rotation."

260. We shall now return to the theorem which suggested the theory of foci of curves of the second order. In the first place let us define :

The axis f of any sheaf of planes of the first order in which planes perpendicular to each other are conjugate with respect to a cone of the second order shall be called a *focal axis* of the cone.

If a plane rotates about a focal axis f its conjugate normal plane at the same time describes the sheaf f, and this axis must consequently (Art. 257) lie in a plane of symmetry for the cone. As is easily seen, a principal axis is at the same time a focal axis only if the cone is a cone of revolution and the principal axis its axis of rotation. Cones of revolution, which we shall hereafter exclude from our discussions, have moreover only one focal axis, namely, the axis of rotation.

The plane of two real focal axes f and f' is a plane of symmetry for the cone, since it is conjugate to the two planes through f and f' which are normal to it. No real plane tangent to the cone can be passed through a focal axis (compare Art. 227).

261. A plane of symmetry a for a cone of the second order is cut by any two conjugate normal planes in two rays s and s' such that every plane passing through s is normal and conjugate to one passing through s' (Art. 257). The two sheaves of conjugate normal planes s and s' are projectively related, and are cut by a second plane of symmetry β in two projective sheaves of rays. These are in involution since any two homologous rays stand to each other in the same relation as do s and s', and the double rays of this involution are focal axes of the cone. If the double rays are imaginary, there are two rays from which the involution β is projected by sheaves in which corresponding planes are at right angles (compare Art. 229) and which are real focal axes of the cone. The cone cannot have more than two real focal axes since the plane of two real focal axes is always a plane of symmetry and a plane of symmetry cannot contain more than two focal axes, and also since a focal axis is also a principal axis only for the cone of revolution.

262. Any two conjugate normal planes are harmonically separated by the two real focal axes f and f' of a cone of the second order : for two such planes intersect the plane of symmetry ff' in two conjugate rays of the involution, of which f and f' are the double rays. In particular, the other two planes of symmetry are harmonically separated by f and f' ; consequently, the angles formed by f and f' are bisected by two principal axes.

We can now distinguish the three principal axes as follows : the first is perpendicular to the plane ff' and lies outside the cone ; the second lies inside, the third outside the cone, and both in the plane ff'. The first of the three planes of symmetry contains the two real focal axes, the second lies wholly outside the cone, while the third, like the first, cuts the cone in two real rays through the vertex ; the second and third planes of symmetry each contain two conjugate imaginary focal axes.

263. The planes bisecting the angles which are formed by two tangent planes to the cone are conjugate normal planes, and as such are harmonically separated by the two focal axes f and f'; they therefore bisect the angles made by the two planes which are determined by the line of intersection t of the tangent planes and the focal axes f and f'.

Similarly we may prove the theorem (comp. Art. 223):

"Every tangent plane of a cone of the second order forms "equal angles with the two planes determined by its ray of contact "and the focal axes of the cone. Two confocal cones of the second "order make right angles with each other along their common "rays."

264. If g is the inverse ray of f with respect to one of two tangent planes which intersect in a ray t, and g' the inverse of f' with respect to the other, the planes tg and tf' form the same angle with each other as do the planes tf and tg', and since the angle tg equals the angle tf and the angle tf' equals the angle tg', the three-edges gtf' and ftg' are congruent and may be brought into coincidence by a rotation about their common edge t; the face angles gf' and fg' are therefore equal; only one of these angles however alters its position as one of the two tangent planes glides around the cone, hence they are of constant magnitude.

From this it follows that—

"The inverse rays of a focal axis f with respect to the tangent "planes of a cone of the second order lie upon a cone of revolution "which has the other focal axis for axis of rotation."

In accordance with the theorem of Art. 263, each of these inverse rays must lie in a plane with the other focal axis and the ray of contact of the corresponding tangent plane; and since the ray of contact forms the same angle with the first focal axis as with its inverse ray, we have—

"The sum, or the difference, of the two angles which an arbitrary "ray of the cone makes with the two focal axes f and f', is constant."

A constant sum or a constant difference is obtained according as we use the one or the other of the supplementary angles which a ray makes with one of the focal axes. Hence it follows that the cone encloses a greater angle of its first plane of symmetry than of the third.

265. The cone is cut by a plane perpendicular to the focal axis f in a curve of the second order of which one focus lies upon f; for any two rays through this point, which are conjugate with respect to the curve, intersect at right angles, since they lie in two conjugate normal planes through f. We can therefore transfer to the cone without further demonstration two previously demonstrated theorems (Arts. 236 and 239) in the following forms:

"If planes be passed through a focal axis and each of the rays "of contact and the line of intersection of two tangent planes, the "plane through the line of intersection makes equal angles with "those through the rays of contact."

"The projective sheaves of rays in which two tangent planes "of a cone of the second order are cut by the remaining tangent "planes are projected from either focal axis of the surface by two "equal and directly projective sheaves of planes."

Let k^2 be the curve in which the cone is cut by a plane normal to f, t a tangent, and F the focus lying upon f. A plane passed through f at right angles to t cuts t in a point of the circle which touches k^2 at the extremities of its major axis (Art. 237); at the same time it cuts the tangent plane of the cone passing through t in the orthogonal projection of f. Hence:

"If the focal axes f, f', of a cone of the second order k^2 be "projected orthogonally upon the tangent planes, all these projec-"tions lie upon a cone of the second order which touches k^2 along "the two vertex rays of the plane ff', and is cut by a plane "normal to either f or f' in a circle."

266. We shall turn now to the 'cyclic planes' of a cone of the second order, which in a certain sense are reciprocal to the focal axes and may be defined as follows:

The plane of any sheaf of rays of the first order, in which rays perpendicular to each other are conjugate with respect to a cone of the second order, is called a *cyclic plane* of the cone.[*]

The name 'cyclic' is applied to such a plane, since every section of the cone parallel to it is a circle; namely, the section is a curve of the second order whose centre lies upon the pole-ray of the plane, and all pairs of conjugate diameters of the curve are at right angles, since they are parallel to two conjugate rays in the plane. A plane of symmetry is therefore a cyclic plane only for a cone of revolution in which the axis of rotation is the principal axis

---

[*] The plane passes through the vertex of the cone and conjugate rays lie in conjugate planes.—H.

normal to the plane of symmetry. The cone of revolution, which we shall again exclude, has only this one cyclic plane.

267. A cyclic plane can have no real rays in common with the cone, since no real ray of the plane is self-conjugate. Two cyclic planes intersect in a principal axis, namely, in a straight line to which in each of the planes a ray is conjugate and normal.

If we denote in general two rays of the bundle to which the cone belongs as conjugate normal rays when they are normal to each other and conjugate with respect to the cone, then it is easy to see that to any ray l of the bundle (the three principal axes excepted) only one conjugate normal ray l' exists; in l', namely, the polar plane and the normal plane of l intersect.

If now the ray l describes in the bundle an arbitrary sheaf of rays ϵ, its polar and its normal planes describe two sheaves of planes projective to ϵ, and hence :

"If a ray l of the bundle in which a cone of the second order "lies, moves in a plane ϵ, its conjugate normal ray l' describes "in general a cone of the second order of which the pole-ray and "the normal ray of ϵ are elements, as are also the three principal "axes of the given cone. The conjugate normal describes a sheaf "of the first order ϵ' only when ϵ passes through one of the three "principal axes, and in this case ϵ' also contains this principal "axis."

In this special case the two sheaves of planes projective to ϵ are perspective to each other. The relation between the conjugate normal rays of the bundle is an involution of the second order which can be used, just as was that between the conjugate normal planes, to determine the principal axes of the cone. This involution determines all cyclic planes of the cone, *i.e.*, all those planes ϵ which coincide with their corresponding planes ϵ'.

268. The planes through a principal axis a are correlated two and two, so that two conjugate normal rays l, l', always lie in two corresponding planes ϵ, ϵ', of a; the two sheaves ϵ, ϵ', of conjugate normal rays, obtained as above, are clearly projectively related to each other. These sheaves of rays are projected from a second principal axis b by two projective sheaves of planes which are in involution, for any two homologous planes stand in the same relation to each other as do the planes ϵ and ϵ'. The two double planes of this involution b are cyclic planes for the given cone, as is easily seen without further explanation.

Thus through each of the three principal axes there are two cyclic planes; these are however conjugate imaginaries for two of the principal axes, and it is only in the third principal axis that two real cyclic planes k and k' intersect (compare Art. 262). For if there were more than two, say three, real cyclic planes, they would intersect two and two in the three principal axes, and so coincide with the planes of symmetry, which has already been shown impossible. Thus—

"There are two sheaves of parallel planes which cut a cone of "the second order in circles; each of these sheaves contains one "of the two real cyclic planes of the cone."

269. Two principal axes and in general all pairs of conjugate normal rays are harmonically separated by the cyclic planes k and k'; for they lie in pairs of conjugate planes of the involution of which k and k' are the double elements. Of the three planes of symmetry for the cone, two therefore bisect the dihedral angles formed by the two cyclic planes k and k', the third is normal to these planes.

270. Since the lines which bisect the angles formed by any two rays of the cone are conjugate normal rays, and hence are harmonically separated by k and k', we have this theorem:

"If the plane of any two rays of a cone intersects two cyclic "planes, the angle which the one line of intersection makes with "one ray equals the angle which the other line of intersection "makes with the other ray."

Similarly, it is easily shown that—

"The plane angle which the cyclic planes determine upon any "tangent plane of the cone is bisected by the ray of contact. Two "concyclic cones therefore touch each of their common tangent "planes along rays at right angles to each other."

Two conjugate normal rays are harmonically separated by the tangent planes along any two rays of the cone which lie in a plane with one of them. From this fact we can without difficulty deduce the following:

"If a cyclic plane intersects two tangent planes and the plane "of their rays of contact, the line of intersection with the latter "forms equal angles with the lines of intersection with the two "former."

If now we introduce a third variable tangent plane (analogous to Art. 239) we can further conclude that—

"The planes by which a variable ray of a cone of the second "order is projected from two fixed rays of the cone determine "angles of constant magnitude upon each of the two real cyclic "planes. Two projective sheaves of planes which generate the cone "are consequently cut by each cyclic plane in two equal projective "sheaves of rays."

271. Most of these theorems upon cyclic planes and many others may be deduced directly from analogous theorems upon focal axes, if we bring to our aid bundles of rays related orthogonally to each other. Two bundles S and S_1 are said to be related orthogonally to each other if every plane of one is correlated to its normal ray of the other, and thus to each sheaf of planes in one a sheaf of rays projective to it in the other, the plane of the latter sheaf being normal to the axis of the former. Four harmonic planes of one bundle correspond therefore always to four harmonic rays of the other which are respectively normal to these planes. Two elements of the bundle S form the same angle with each other as do their corresponding elements of the bundle S_1.

To the rays of a cone of the second order in the bundle S correspond then the tangent planes to a cone of the second order in S_1; if we look upon the former as being generated by two projective sheaves of planes, the latter will appear as being generated by the corresponding two projective sheaves of rays. Moreover, to any two rays which are harmonically separated by elements of the one cone correspond two planes which are harmonically separated by two tangent planes of the other cone; from which it follows that in general conjugate elements correspond to conjugate elements.

Two rays or planes normal to each other, which are conjugate with respect to one cone, correspond therefore to two planes or rays normal to each other, which are conjugate with respect to the other cone. Each cyclic plane of the one cone consequently corresponds to a focal axis of the other, and likewise each plane of symmetry to a principal axis. We can therefore transfer the properties of focal axes without further comment to cyclic planes, and thus obtain, for example, the theorem (compare Art. 264):

"The sum or difference of the angles which an arbitrary tangent "plane of the cone makes with the two cyclic planes is constant."

272. If there are given two cones of the second order correlated orthogonally to each other, to every plane through the vertex, but

containing no ray, of the one cone there corresponds a ray normal to it, which passes through the vertex and lies within the other cone. If of all plane angles which lie within the second cone that is greatest in which the real focal axes f, f', lie, then of all dihedral angles which lie without the first cone that is greatest within which its cyclic planes lie. Since now, in fact, a cone of the second order does define a greater angle upon its first plane of symmetry ff' than upon its third (Art. 264), we learn that—

"The cyclic planes k and k' of a cone of the second order "intersect in the third principal axis; this lies in a plane with "the second principal axis and the two focal axes, but outside "the cone."

The first two planes of symmetry of the cone therefore bisect the dihedral angles formed by k and k', while the third plane of symmetry is normal to k and k'.

273. A cone of the second order and a sphere concentric with it have in common a 'sphero-conic section.' The sphere cuts each plane of symmetry of the cone in an 'axis' of the sphero-conic, each cyclic plane in a 'cyclic line,' each tangent plane in a 'tangent,' each principal axis in two 'centres,' and each focal axis of the cone in two 'foci' of the sphero-conic.

The sphero-conic is a twin curve, that is, it consists of two separate equal lines whose points lie two and two diametrically opposite upon the sphere. In general it has three pairs of centres in which its three axes intersect at right angles, also two real cyclic lines and two pairs of real foci. The first axis passes through the four foci and the two centres in which the cyclic lines intersect; the second axis lies wholly outside the conic, and has these two centres in common with the first; the third axis intersects the two cyclic lines perpendicularly, and like the first contains two pairs of real vertices of the conic. In a particular case the sphero-conic consists of two equal sphero-circles.

274. All those properties of the cone of the second order which refer to planes of symmetry, principal axes, focal axes, and cyclic planes, can be transferred directly to the sphero-conic section. By this means we obtain among others the following theorems:

"If one vertex of a spherical triangle moves upon the sphere "so that the perimeter of the triangle remains constant, it describes "a sphero-conic section of which the other two vertices are foci."

" If one side of a spherical triangle moves so that the area of the "triangle remains constant, it envelops a sphero-conic section of "which the other two sides of the triangle are the cyclic lines."

EXAMPLES.

1. If a given dihedral angle is rotated about its edge s, the plane of the two lines in which the faces of the given angle are cut respectively by two fixed planes passing through a point of s envelops a cone of the second order touched by the fixed planes, of which s is a focal axis (Art. 265). State the reciprocal theorem.

2. The angles which the tangent planes normal to the plane of the real focal axes f and f' define upon the remaining tangent planes are projected from the focal axes f and f' by right dihedral angles.

3. The real cyclic planes k and k' of a cone of the second order intersect every dihedral angle whose faces are determined by the two rays of the cone normal to the intersection of k and k' and any third ray of the cone, in a right angle.

4. All dihedral angles whose faces are tangent to one or another of a system of confocal cones and pass through a fixed point P are bisected by two fixed planes at right angles.

5. Besides the cone of revolution, special mention should be made of the following cones of the second order :

(a) The equilateral cone (Schröeter) to which right-angled three-edges can be inscribed.

(b) That to which right-angled three-edges can be circumscribed.

(c) The orthogonal cone (Schröeter) whose cyclic planes are normal to two rays of the cone.

(d) That whose focal axes are normal to two tangent planes.

(e) Pappus' cone, which is cut by the planes through two special lines in rays at right angles to each other (the special lines are the pole-rays of the cyclic planes).

(f) Hachette's cone, for which the tangent planes passed through two special rays are normal to each other.

(g) and (h) Those whose focal axes or whose cyclic planes are normal to each other.

6. The lines bisecting the supplementary angles formed by two straight lines, one of which g remains fixed while the other moves in a plane γ, generate a Pappus' cone. The plane γ is a cyclic plane of the cone and g its pole-ray.

From a point P of a sphere the great circles of the sphere are projected by Pappus' cones.

APPENDIX.

THE PRINCIPLE OF RECIPROCAL RADII.

§ 1. Two points P and P_1 are said to be 'inverse' with respect to a circle of radius r and with centre M, if they lie in a straight line with M and are conjugate with respect to the circle. They are harmonically separated by the extremities of the diameter through them, and consequently (Art. 72)

$$MP \cdot MP_1 = r^2 \text{ or } MP = \frac{r^2}{MP_1}.$$

The product of the radii vectores of two inverse points is thus constant, or the radius vector of any point P is inversely proportional to that of its inverse point P_1. In consequence, this phase of 'inversion' has obtained* the name 'The Principle of Reciprocal Radii'; M is called the 'inversion centre,' and r^2 the 'power' of the reciprocal radii.

To every point within the curve a point outside the curve is inverse, and to every infinitely distant point of the plane the centre is inverse; the points of the circle are inverse to themselves.

§ 2. The polar of a point P with respect to the given circle is perpendicular to MP at the inverse point P_1. The inverse points P_1, Q_1, R_1 ... of any given points P, Q, R ... of a straight line g are found by dropping perpendiculars from G the pole of g upon the straight lines MP, MQ, MR ... respectively. Hence

"The inverse of any straight line g is a circle γ the extremities "of whose diameter normal to g are the centre M and the pole G "of the given line."

Conversely, the inverse of a circle passing through M is a straight

---

* Liouville in the *Journal de Mathématiques*, I. Série, T. XII., p. 265.

line g; for, to the straight line joining two points A and B whose inverse points A_1 and B_1 lie upon γ a circle is inverse, which passes through A_1, B_1, and M and is consequently identical with γ. On account of this characteristic Moebius* called this quadratic transformation a 'circular transformation.'

§ 3. If through the inversion centre M we draw a parallel to the straight line g, this will touch the circle γ inverse to g at M, since it is perpendicular to the diameter MG of γ.

From this we conclude :

"Any two straight lines of the plane f and g intersect at the "same angle as do their inverse circles ϕ and γ."

Two indefinitely small inverse triangles have therefore equal angles and are similar; or in other words,

"By inversion the plane is transformed into itself so that the "infinitesimal portions of two inverse figures are similar."

This theorem is applicable at all points of the plane except for the indefinitely small portion which lies about the centre M.

§ 4. Since a straight line can be looked upon as a circle of infinite radius, we may consider the theorem "to every circle passing through M a straight line is inverse," and its reciprocal (§ 2) as special cases of the following theorem :

"To every circle k a circle k_1 is inverse; M is a centre of "similitude for k and k_1."

In order to prove this general theorem we choose upon k a fixed point P and a movable point Q and mark their two inverse points with P_1 and Q_1, and the points in which k is intersected by the secants MP and MQ a second time, with P' and Q', respectively. Then, from the nature of the segments of secants of circles,

$$MP . MP' = MQ . MQ',$$

and from the principle of reciprocal radii,

$$MP . MP_1 = MQ . MQ_1 (= r^2);$$

consequently, $$MP' : MP_1 = MQ' : MQ_1$$

and the triangles $MP'Q'$ and MP_1Q_1 are similar.

If then Q, and at the same time Q', traverses the circle k, Q_1 describes a curve similar and similarly situated to k, that is, describes a circle k_1. Evidently M is a centre of similitude, while

---

* Moebius in the *Abhandlungen der Königl. Sächs. Gesellschaft der Wissenschaften*, Leipzig, 1855, Bd. II., S. 531-595, and in the reports of the proceedings of the same Society for 1853, pp. 14-24.

Q' and Q_1, also P' and P_1, are 'homologous' points of the two circles k and k_1.

§ 5. "Any circle which passes through two inverse points P and P_1 "is inverse to itself and intersects the assumed circle whose points "coincide with their inverse points orthogonally."

For it has in common with the circle inverse to it the points P and P_1 inverse to each other and two points inverse to themselves in which it is touched by straight lines through M, since $MP . MP_1 = r^2$.

§ 6. If we assume two systems of points P, Q, R ... and P_1, Q_1, R_1 ... inverse to each other in a plane, and one system rotates about the centre M till each of its points has described a semi-circle, the pairs of inverse points P and P_1, Q and Q_1, etc. are again in straight lines with M, but now lie upon opposite sides of M. As before,

$$MP . MP_1 = MQ . MQ_1 = MR . MR_1 = \ldots = \text{a constant};$$

but the 'power,' that is, the product of the radii vectores of inverse points, has no longer a positive value but is now negative.

Thus we obtain a second case of inversion which is distinguished from the first by having no point of the plane coincident with its inverse point. In this second case also the plane is transformed into itself conformally; to each line a circle through M is inverse, and in general, to each circle k a circle k_1; here again the centre of reciprocal radii is a centre of similitude of two inverse circles.

§ 7. The points of space can likewise be coördinated two and two so that each is the inverse of the other, the centre M and a positive or negative value of the power being chosen at pleasure. This can be accomplished most easily by first coördinating to each other, two and two, the points of any plane through M and then rotating this plane about some axis through M. Two inverse points of the plane indicate in each position of the plane two inverse points of space.

§ 8. If we rotate the plane about the line of centres of two inverse circles, which passes through M, the circles will describe two spheres; hence,

"The inverse of any sphere k is a sphere k_1; M is a centre of "similitude of k and k_1. Any plane γ is inverse to a sphere which "passes through M and which is touched in M by a plane parallel "to γ."

o

The last part of this theorem may be looked upon as a special case of the first part, or it may easily be proved independently. And further,

"Any two planes in space intersect at the same angle as do "the spheres inverse to them."

Two indefinitely small tetrahedra which are inverse to each other have therefore equal dihedral angles and hence equal face angles; they are, as a little reflection will show, similar if the power of the reciprocal radii is negative, and symmetrical if the power is positive. Since, therefore, their homologous faces are always similar,

"Two inverse surfaces are depicted conformally upon each "other."

§ 9. In order, then, to depict a sphere k conformally upon a plane s, choose as centre of inversion M one of the two points of k whose tangent planes are parallel to s and fix the power equal to the product of the two segments MP and MP_1 which k and s determine upon any straight line passed through M. The plane s is thus inverse to a sphere which has in common with the given sphere k the points P and M and the tangent plane at M, and which consequently coincides with k. Hence :

"If a sphere k is projected (stereographically) from one of its "points M upon a plane s which is parallel to the tangent plane at "M, the surface k will be depicted conformally upon the plane s."

Much use is made of this stereographic projection, which was known as early as the time of the astronomer Ptolemy, in the construction of geographical maps. By this means it is ensured that the most minute angles upon the map have the same size as the angles corresponding to them upon a globe ; the lengths of different lines upon the globe are always represented upon the map according to a variable scale, since a sphere cannot be flattened out upon a plane without distortion.

§ 10. The inverse of any circle is always a circle ; two spheres intersect in the latter, which are inverse to two spheres through the former. If, in particular, the one circle passes through M, the other becomes a straight line. The meridians and parallels of latitude upon a globe transform therefore by stereographic projection into two systems of orthogonal circles ; the projections of the meridians are circles which intersect in two points (the north and south poles), and to these the projections of the parallels of latitude, which have no real point in common, are at right angles.

Only the circles of the sphere which pass through M are represented upon the plane of projection by straight lines.

If the centre of projection M is placed at the north or south pole, the parallels of latitude will be depicted as a system of concentric circles and the meridians as their diameters.

§ 11. For any three given spheres there can in general be found a circle which intersects all three at right angles; it lies in the plane of centres and its centre is the radical centre of the three circular sections. The construction for this circle is only impossible when the spheres have either one or two points in common.

If now we choose a point of this orthogonal circle as centre of inversion, the circle itself is transformed into a straight line and the spheres into three spheres which are cut orthogonally by this straight line and whose centres consequently lie upon the line. Three spheres can therefore always be transformed by inversion into three others whose centres lie in one straight line, except when the three given spheres have a point in common, in which case they can be transformed into three planes.

§ 12. A system of spheres which is described by the continuous motion of a variable sphere envelops in general a surface F which has a system of circular lines of curvature. Each sphere of the generating system touches the surface F along the circle which it has in common with the next consecutive sphere of the system; and since the normals to F in the points of this line intersect in the centre of the sphere, the circle is a line of curvature of F. If now the surface F is transformed by inversion into another surface F_1, these lines of curvature become circular lines of curvature of F_1, since F_1 is enveloped by that system of spheres which is inverse to the system enveloping F. All surfaces inverse to surfaces of revolution have therefore a system of circular lines of curvature.

§ 13. One of the most remarkable of these surfaces is the 'cyclide' discovered by Dupin. This is generated by a variable sphere which constantly touches three given spheres. The theory of this cyclide can be developed easily upon the basis of the preceding paragraphs, at the same time proving the following statements:

A Dupin's cyclide is always transformed by inversion into a cyclide; if the centre of inversion is properly chosen it can be transformed into a rotation cyclide which is generated by a sphere or a plane rotating about an axis. The cyclide therefore comprises two systems of circular lines of curvature and is touched along these by two systems of spheres; the centres of these spheres and lines of curvature lie in two planes of symmetry of the cyclide, which are normal to each other and in each of which two lines of curvature lie. Every sphere of the one system touches all spheres of the other system along the points of a circular line of curvature. A sphere can be passed through two lines of curvature if they belong to the same system; if to opposite systems they intersect in one point at right angles.

§ 14. The cyclide has either no (real) double point, or two conical-points in which all lines of curvature of one system intersect, or a cuspidal point in which all lines of curvature of one system are tangent to each other. Essentially different forms of these three principal varieties are obtained if we transform the corresponding rotation cyclide by reciprocal radii whose centre is chosen (1) inside, (2) upon, and (3) outside the rotation cyclide. The last two principal varieties can be depicted conformally by inversion upon a right cone and a cylinder, respectively.

The planes in which the lines of curvature of either system lie, two and two, all intersect in one straight line. This lies in the one plane of symmetry, and is perpendicular to the other. Each of two planes touches the surface along the points of a circle.

If spheres be passed through the lines of curvature of either system and a fixed point M, these will all intersect in one circle.

If a cyclide extends to infinity, as may happen in any one of the three varieties, it must have two right lines of curvature which are perpendicular to each other. The planes of all remaining lines of curvature pass partly through one and partly through the other of these two gauche lines.

RULED SURFACES OF THE THIRD ORDER.

§ 15. A range of points u of the first order and a range k^2 of the second order which are projective, but neither lie in the same plane nor have a self-corresponding point, generate a system of straight lines which we shall call a 'regulus of the third order.' Any straight line g intersects at least one and at most three lines of this system, as is easily seen (Art. 194) if we project the range u from the axis g by a sheaf of planes. The regulus of the third order is in general projected from an arbitrary point by a sheaf of planes of the third order.

§ 16. Every plane passing through u which cuts the curve k^2 contains two rays of the regulus. If now from D, the intersection of two such rays, we project the two ranges of points u and k^2, we obtain a sheaf of rays of the first order and a cone of the second order which are projectively related and have two rays self-corresponding, and which generate (Art. 195) a sheaf of planes d of the first order. All pairs of rays of the regulus which lie in a plane through u intersect therefore in the points of a straight line d, and from each of these points of intersection the regulus is projected by a sheaf of planes of the first order perspective to the ranges u and k^2. We may call the straight lines u and d ' directors' of the regulus, since they intersect all rays of the system. The straight line d coincides with u only when u intersects the curve k^2; in this special case the director u is also a ray of the regulus.

§ 17. From a point S which lies upon a ray a of the regulus, but upon neither director, the ranges u and k^2 are projected by a sheaf of the first order and a cone of the second order which have the ray a self-corresponding, and consequently (Art. 197) generate a sheaf of planes of the second order perspective to u and k^2. The regulus, therefore, is projected from any point S lying upon it by a sheaf of planes of the second order. Since this sheaf is projective to the sheaf of planes d, which in turn is perspective to the ranges u and k^2,

§ 18. The regulus of the third order may be generated by a sheaf of planes d of the first order, upon whose axis the rays of the system intersect, two and two, and a sheaf of planes S of the second order projective to the sheaf d, whose centre may be chosen anywhere upon the regulus.

This second method of generating the regulus is reciprocal to the method first stated, and the regulus of the third order is consequently self-reciprocal.

From the second method too we obtain properties which are reciprocal to those just deduced, among others :

The regulus of the third order is projected from any point which lies upon one of its rays by a sheaf of planes of the second order ;	The regulus of the third order is intersected by any plane which passes through one of its rays in a curve of the second order ;

and all such projecting sheaves and curves of section are projective to one another, to the range of points u and to the sheaf of
planes d. The curves of section are perspective to the sheaf d
and to all sheaves of planes of the second order projecting the
regulus.

§ 19. The regulus of the third order lies upon a ruled surface
of the third order F^3 which is intersected by any straight line in
at most three points, and is cut by any plane in a curve of the
third order. The surface F^3 passes twice through the director d,
and of any plane section of the surface either an actual or
an isolated double point lies upon this director. Planes passed
through a ray of the regulus intersect the ruled surface in that
ray and a curve of the second order; the curve is intersected
by the ray in two points, one of which lies upon the double line d,
while at the other the plane of section is tangent to the surface.
All planes passing through the rays of the regulus are thus tangent
to the surface of the third order; the planes through the director u,
in each of which two rays of the surface lie, are doubly-tangent
planes of the surface.

§ 20. Three varieties of the ruled surface of the third order
may be distinguished. Any one of the curves k^2 of the second
order which lie upon the surface F^3 may either enclose the point
in which its plane cuts the axis u of doubly-tangent planes, or this
point may lie outside the curve, or it may lie upon the curve; and
whatever be the relation between this one curve and point, the
same relation will exist between any other such curve and
corresponding point lying upon the same surface.

In the first case (as is seen from the method of generating the
surface given in § 15) two rays of the regulus lie in each plane
through u, and at the same time intersect in a point of d. In
the second case the so-called cuspidal planes of the surface, which
pass through u and are tangent to the curve k^2, separate the
actual doubly-tangent planes and the actual double points of the
surface from the 'isolated' doubly-tangent planes which pass
through u and the isolated double points which lie upon d. The
third case is to be looked upon as the limiting case between the
first two; here the two directors u and d (and also the two cuspidal
planes) coincide.

§ 20a. Suppose there are given two gauche lines u and d and a
conic section k^2 which is met in a point by the one straight line d

but which lies in a plane with neither of the lines, and let a straight line g move so as constantly to intersect u, d, and k^2; this line will describe a ruled surface of the third order of which d is a double line and whose doubly-tangent planes pass through u. State the reciprocal of this theorem.

QUADRANGLES AND QUADRILATERALS WHICH ARE SELF-POLAR WITH RESPECT TO CONIC SECTIONS.

§ 21. A complete quadrangle is called a 'self-polar quadrangle' with respect to a conic section γ^2 if each of its six sides is conjugate with respect to γ^2 to its opposite side. Similarly, a complete quadrilateral is called a 'self-polar quadrilateral' if each of its six vertices is conjugate to its opposite vertex. The polars of the vertices of a self-polar quadrangle form a self-polar quadrilateral; the six sides of the quadrangle pass through the six vertices of the quadrilateral.

§ 22. If two pairs of opposite sides of a complete quadrangle consist of conjugate rays with respect to a conic γ^2, the same is true of the third pair, and the quadrangle is self-polar with respect to γ^2. For the polar of any vertex intersects the quadrangle in an involution (Art. 219) in which these two pairs of opposite sides, and hence all three pairs, pass through conjugate points.

The reciprocal theorem, viz., a complete quadrilateral is self-polar if two pairs of opposite vertices are conjugate with respect to a conic, is proved in a similar manner.

§ 23. If the vertices A, B, C, of a triangle be joined to the poles of their opposite sides, the three joining lines will pass through one point D which taken with A, B and C forms a self-polar quadrangle with respect to the conic γ^2 (§ 22). Three vertices A, B, C, of a self-polar quadrangle, chosen arbitrarily, thus determine the fourth vertex D; in particular, if A and B are conjugate with respect to γ^2 they form with D a self-polar triangle of γ^2. A quadrangle which consists of a self-polar triangle and any fourth point of the plane is a self-polar quadrangle with respect to γ^2 (§ 21). Improper self-polar quadrangles of which three vertices lie in one straight line or two vertices coincide may also arise.

§ 24. If the three sides a, b, c, of a triangle are made to intersect the polars of their opposite vertices, the three points of intersection will lie upon one straight line d, which taken with a, b and c forms a self-polar quadrilateral of the conic (§ 22). Three sides of a self-polar quadrilateral thus determine the fourth side; it is only when the three sides form a self-polar triangle that the fourth side can be chosen arbitrarily.

§ 25. If two self-polar quadrangles *ABCD* and *ABC'D'* of a conic γ^2 have two vertices *A* and *B* in common, their six vertices lie upon a curve of the second order, which may however degenerate into the two straight lines *AB* and *CD*.

If two self-polar quadrilaterals of a conic γ^2 have two sides in common, their six sides are tangent to a curve of the second class, *i.e.*, they belong to a sheaf of rays of the second order, which may however degenerate into two points.

Since the sheaves of rays *A* and *B* are related projectively to each other when to each ray of *A* is correlated its conjugate ray of B, then

$$A(CDC'D') \barwedge B(DCD'C'),$$

and hence (Art. 215)

$$A(CDC'D') \barwedge B(CDC'D'),$$

as is asserted in the theorem on the left.

§ 26. Since a self-polar triangle of γ^2 is converted into a self-polar quadrangle by the addition of any point of the plane, we derive the following theorems from those just stated:

If a self-polar triangle and a self-polar quadrangle have one vertex in common, their six vertices lie upon a curve of the second order.

If a self-polar trilateral and a self-polar quadrilateral have one side in common, their six sides are tangent to a curve of the second class.

Any two self-polar triangles of a conic γ^2 can be inscribed in a curve of the second order and circumscribed to a curve of the second class (compare Art. 215, note). But these curves may degenerate into two straight lines and two points, respectively.

§ 27. Two conics γ^2 and γ^2_1 which lie in the same plane have an infinite number of self-polar quadrangles and self-polar quadrilaterals in common. Two sides *a* and *b* of any such quadrangle may be chosen at random. The opposite sides a_1 and b_1 join their two poles with respect to the given conics γ^2 and γ^2_1.

If two self-polar quadrangles *ABCD* and *AB'C'D'* common to γ^2 and γ^2_1 have one vertex *A* in common, their seven vertices lie upon a curve of the second order.

If two self-polar quadrilaterals common to γ^2 and γ^2_1 have one side in common, their seven sides touch a curve of the second class.

If, namely, $ABB'P$ is a self-polar quadrangle of γ^2, and $ABB'Q$, a self-polar quadrangle of γ^2_1, the conics $ABCDB'P$ and $ABCDB'Q$ have five points in common and therefore coincide with the conic $ABB'PQ$; in the same way, the conics $AB'C'D'BP$ and $AB'C'D'BQ$ coincide with $ABB'PQ$.

§ **28.** With respect to three given conics γ^2, γ^2_1, γ^2_2, of a plane, a straight line a has three poles which in general do not lie upon one straight line. If a describes a sheaf of rays U, its poles describe three ranges of points u, u_1, u_2, projective to U and hence to each other; of the rays which join homologous points of u and u_1, at most three and at least one passes through the corresponding point of u_2 (Art. 192) and is conjugate to a ray of U with respect to all three of the conics γ^2, γ^2_1, γ^2_2. From this we derive the first of the following reciprocal theorems:

In a plane there is an infinite number of pairs of rays which are conjugate with respect to three given conic sections. These pairs of rays in general envelop a curve of the third class, and any two of them form two pairs of opposite sides of a self-polar quadrangle common to the three given conics (§ 22). Three tangents which can be drawn from a point U to the curve of the third class form with their three conjugate tangents the three pairs of opposite sides of a self-polar quadrangle common to the three conic sections.

In a plane there is an infinite number of pairs of points which are conjugate with respect to three given conic sections. The locus of these pairs of points is in general a curve of the third order, and any two of them form two pairs of opposite vertices of a self-polar quadrilateral common to the three conics. Three points in which any one straight line intersects the curve of the third order form with the points of the curve conjugate to them the three pairs of opposite vertices of a self-polar quadrilateral common to the three conic sections.

NETS AND WEBS OF CONIC SECTIONS.

§ **29.** If a conic section k^2 is circumscribed to a quadrangle which is self-polar with respect to another conic section γ^2, the two curves γ^2 and k^2 have the following remarkable relations to each other:

(a) Any three points of k^2 determine a self-polar quadrangle of γ^2 (§ 23) whose fourth vertex also lies upon k^2.

(a_1) Any three tangents to γ^2 determine a self-polar quadrilateral of k^2 whose fourth side is also a tangent to γ^2.

(*b*) If k^2 is intersected by two straight lines which are conjugate with respect to γ^2, the points of intersection are the vertices of a quadrangle which is self-polar with respect to γ^2.

(*c*) To the curve k^2 there can therefore be inscribed an infinite number of quadrangles and an infinite number of *real* triangles which are self-polar with respect to γ^2; every point of k^2 enclosed by γ^2 is a vertex of one of these self-polar triangles.

(*d*) The polar of a point of k^2 with respect to γ^2 intersects either one or both of the curves in two real points which are conjugate with respect to the curve upon which they do not lie.

(*e*) Two tangents to γ^2, whose points of contact are conjugate with respect to k^2, always intersect in a point of k^2.

(*b₁*) If tangents be drawn to γ^z from two points which are conjugate with respect to k^2, the four lines so drawn are the sides of a quadrilateral which is self-polar with respect to k^2.

(*c₁*) To the curve γ^2 there can therefore be circumscribed an infinite number of quadrilaterals, and in general an infinite number of *real* triangles which are self-polar with respect to k^2; every tangent to γ^2 which lies wholly outside k^2 is a side of one of these self-polar triangles.

(*d₁*) From the pole with respect to k^2, of a tangent to γ^2, there can be drawn to one or to both curves two real tangents which are conjugate with respect to the conic not touched by them.

(*e₁*) Two points of k^2, the tangents at which are conjugate with respect to γ^2, always lie upon the same tangent to γ^2.

(*f* and *f₁*) The curve k^2 lies either wholly outside γ^2 or partly outside and partly inside; neither of the two curves wholly encloses the other.

§ 30. The following will be sufficient proof of these theorems:

Two vertices of the self-polar quadrangle of γ^2 to which the conic k^2 is circumscribed determine, with any third point A of k^2, a second self-polar quadrangle of γ^2 which is inscribed to k^2 (§ 23 and § 25); from this second self-polar quadrangle a third can be derived in the same way, of which two vertices A and B are chosen arbitrarily upon k^2, and this third yields a fourth self-polar quadrangle of γ^2 inscribed to k^2, which has three vertices A, B, C, arbitrarily chosen upon k^2. This proves theorem (*a*), and from it (*b*) follows immediately.

Let a be the polar of A with respect to γ^2, and if this should have no real point in common with γ^2, A lies inside γ^2; the polar a intersects the opposite sides of the self-polar quadrangle of γ^2 which has one vertex at A, in pairs of points conjugate with respect to γ^2 (§ 22), while the involution so determined upon a is elliptic, *i.e.*, has no real double elements. The conic sections which can be circumscribed to any one of these self-polar quadrangles must therefore intersect the straight line a in two real conjugate points which, taken with A, form a real self-polar triangle of γ^2; for, these conics form a continuous series, an infinite number of them have two real conjugate points in common with the involution a (Art. 221), and it is impossible that one of them should have imaginary points of intersection, since then between this conic and those having real intersections there would lie a conic tangent to a at a real double element of the involution. Thus theorems (c) and (f) are proved, and from these (d) and (e) follow easily.

§ 31. Theorems (a_1) to (f_1) of § 29 can be proved in the same way as those numbered from (a) to (f) as soon as it is shown that to the conic γ^2 some one quadrangle can be circumscribed which is self-polar with respect to k^2. In order to show this, choose three points P, Q, R, upon k^2, whose joining lines lie outside γ^2, this being always possible (§ 29f). These determine a quadrangle $PQRS$ self-conjugate with respect to γ^2 and inscribed to k^2, whose opposite sides intersect in three points U, V, W, lying outside γ^2; these three points are conjugate two and two with respect to k^2, and two of them lie outside k^2. If now from U and V pairs of tangents be drawn to γ^2, these harmonically separate the pairs of opposite sides of the self-polar quadrangle $PQRS$ (Art. 143); consequently, these four tangents form a quadrilateral whose opposite vertices are harmonically separated by the pairs of opposite sides of the quadrangle $PQRS$, and since one of the six vertices of the quadrilateral, as is easily seen, lies within k^2, it must also be harmonically separated from its opposite vertex by the curve k^2 (Art. 220). Two opposite vertices of the quadrilateral are thus conjugate with respect to k^2; this quadrilateral circumscribing γ^2 is consequently self-polar with respect to k^2.

§ 32. Of two conic sections k^2 and γ^2, of which the one k^2 is circumscribed to a self-polar quadrangle of the other γ^2, or of which γ^2 is inscribed to a self-polar quadrilateral of k^2, we say* that "the curve of the second order k^2 *supports* or *carries* the

---

* Compare my article in Crelle-Borchardt's *Journal für die reine und angewandte Mathematik*, Bd. 82.

curve of the second class γ^2," or, on the other hand, "γ^2 *is supported by* or *rests upon* k^2." If k^2 breaks up into two straight lines, or γ^2 into two points, we say accordingly,

A line-pair* supports the curve of the second class γ^2 if its rays are conjugate with respect to γ^2.

A point-pair* rests upon the curve of the second order k^2 if its points are conjugate with respect to k^2.

And since a straight line is conjugate to itself with respect to γ^2 only when it touches γ^2, we add,

A two-fold line supports the curve of the second class γ^2 if it is tangent to γ^2.

A two-fold point rests upon the curve of the second order k^2 if it is a point of k^2.

§ 33. All conics which support a given curve of the second class γ^2 and pass through three given points chosen at random, are circumscribed to the self-polar quadrangle of γ^2 which is determined by these points (§ 29a); through an arbitrary fourth point of the plane there passes therefore only one of these conics. Hence :

The conic sections k^2 which support a given curve of the second class γ^2 form a manifold of four dimensions which we shall call a 'net of conics of the fourth grade' (*lineares Kegelschnitt-System vierter Stufe*). About an arbitrary quadrangle there can be circumscribed in general only one conic of this net; for the quadrangle becomes self-polar with respect to γ^2, and all conics circumscribed about it belong to the net, as soon as any two of them support the curve γ^2.

The conic sections γ^2 which rest upon a given curve of the second order k^2 form a manifold of four dimensions which we shall call a 'web of conics of the fourth grade' (*lineares Kegelschnitt-Gewebe vierter Stufe*). To an arbitrary quadrilateral there can be inscribed in general only one conic of this web; for a quadrilateral becomes self-polar with respect to k^2, and all conics inscribed in it belong to the web, as soon as any two of them rest upon the curve k^2.

§ 34. Of the conic sections which support two given curves of the second class γ^2 and γ^2_1, in general only one passes through three given points A, B, C, of the plane. This conic also passes

---

* A curve of the second order which breaks up into two straight lines will be spoken of as a *line-pair*, while a curve of the second class which breaks up into two points will be called a *point-pair*. —H.

through the two points which with *A*, *B*, and *C* form self-polar quadrangles of γ^2 and γ^2_1, respectively. In general then we may say :

The conic sections k^2 which support two, three, or four curves of the second class, chosen arbitrarily in a plane, form a triply, doubly, or singly infinite manifold, which we shall call a net of conics of the third, second, or first grade, as the case may be.

The conic sections γ^2 which are supported by two, three, or four curves of the second class, chosen arbitrarily in a plane, form a triply, doubly, or singly infinite manifold, which we shall call a web of conics of the third, second, or first grade, as the case may be.

For particular relative positions of the given curves these statements are subject to exceptions which are quite apparent. The number indicating the grade does not merely suggest of how many dimensions the net or web of conics is, but it is also, as we shall see (§§ 45 and 51), equal to the number of points or tangents by which one of its conics is determined.

§ 35. The net of conics of the first grade is ordinarily called a 'sheaf of conics,' and whenever the term 'net of conics' is used without specifying the grade it is to be understood that the net is of the *second* grade; the reciprocal forms bear the names 'range of conics,' and 'web of conics,' it being understood in the latter that, as before, the web is of the second grade unless otherwise specified. To be sure, we are accustomed elsewhere to define the 'sheaf of conics' and the 'range of conics' as the totality of conics which 'circumscribe a quadrangle' and 'are inscribed in a quadrilateral,' respectively, but these definitions are contained in the above, as the following theorems will show :

If two conics of a net of any grade are circumscribed to a quadrangle, this quadrangle is self-polar with respect to all curves of the second class which are supported by the net (§ 33), and consequently all conics which circumscribe the quadrangle belong to the net.

If two conics of a web of any grade are inscribed in a quadrilateral, this quadrilateral is self-polar with respect to all curves of the second order which support the web (§ 33), and consequently all conics inscribed in the quadrilateral belong to the web.

§ 36. The theory of nets of conics of the fourth grade is not essentially different from the polar theory of the curve of the second class γ^2 which rests upon all conics of the net. We shall therefore only call attention to the fact that the net is a special one if γ^2 reduces to two points *P* and *Q*, or to a two-fold point *P*. Thus,

| All conics of the plane which pass through a given point P, or with respect to which two points P and Q are conjugate, form a special net of conics of the fourth grade. | All conics of the plane which touch a given straight line, or with respect to which two straight lines are conjugate, form a special web of conics of the fourth grade. |

For example, all parabolas of a plane form a special web of conics of the fourth grade. To what two imaginary points does the curve of the second class γ^2 which rests upon the equilateral hyperbolas of a plane reduce? The conics of a plane, of which one axis has a given direction, form a special net of the fourth grade to which belong all circles of the plane; a circle can be circumscribed about any quadrangle which is common to two of these conics.

NETS AND WEBS OF CONICS OF THE FIRST AND THIRD GRADES.*

§ 37. If there are given two vertices A and B of a self-polar quadrangle of a curve of the second class γ^2, and the side u conjugate to AB, upon which the other two vertices C and D lie, these latter vertices are a pair of coördinated points of an involution upon u. This involution is obtained if to any ray AC or AD of A is correlated the ray BD or BC conjugate to it of B and then the intersections with u of pairs of corresponding rays of the two sheaves A and B thus related projectively to each other are noted. If now we wish to construct a common self-polar quadrangle of two conics γ^2 and γ^2_1, having the points A and B for vertices, the straight line u upon which the other two vertices C and D lie must pass through the two poles of AB with respect to the conics γ^2 and γ^2_1; the points C and D are coördinated to each other in each of two involutions lying upon u, and are therefore uniquely determined (Art. 248), unless, as an exceptional case, the two involutions are identical. The double elements of each of these involutions are conjugate with respect to all conics which can be circumscribed to the self-polar quadrangle $ABCD$; and, if C and D are imaginary there is always one of these conics which passes through any chosen point P, as was shown in Art. 254.

---

* Compare Schröter, *Die Theorie der Kegelschnitte*, II. ed., pp. 224-403 (Leipzig, 1876).

Hence (compare § 34),

<table>
<tr>
<td>

§ 38. All conics which support two curves of the second class γ^2 and γ^2_1 and pass through two real points are circumscribed to a common self-polar quadrangle of γ^2 and γ^2_1, whose remaining two vertices may however be conjugate-imaginary.

</td>
<td>

All conics which rest upon two curves of the second order k^2 and k^2_1 and touch two real lines are inscribed to a common self-polar quadrilateral of k^2 and k^2_1, whose remaining two sides may however be conjugate-imaginary.

</td>
</tr>
</table>

We may consider this self-polar quadrangle fully known (Art. 244) if, besides its two real vertices A and B, there be given any one of its circumscribed conics and the straight line u upon which the other two vertices lie. If two circumscribed conics are given, project all the points of one of them both from A and from B upon the other, and we obtain upon this latter curve two projective ranges of points in which C and D, the remaining vertices of the inscribed quadrangle, are the self-corresponding points; the straight line u is thus determined by a known construction, even if C and D are imaginary (Art. 240). The self-polar quadrangle then is determined as well by two of its circumscribed conics; all conics circumscribing the quadrangle support both γ^2 and γ^2_1.

§ 39. Upon the basis of the two preceding sections and the theorems of § 35 we can now prove the following important theorems, which most intimately connect the net and web of conics of the third grade with those of the first grade.*

<table>
<tr>
<td>

With a net of conics of the third grade there is always connected a range of conics, in such a manner that each curve of the range rests upon every curve of the net.

</td>
<td>

With a web of conics of the third grade there is always connected a sheaf of conics, in such a manner that each curve of the sheaf supports all curves of the web.

</td>
</tr>
</table>

Suppose that γ^2 and γ^2_1 are the two curves of the second class which rest upon all conics of the net of the third grade and determine it (§ 34). Choose four conics k, l, m, n, in the net so that k, l, and m pass through any chosen real point and intersect two and two in the vertices of three different self-polar quadrangles (kl), (km), (lm), common to γ^2 and γ^2_1; the fourth conic n does not pass through P

---

* These theorems originated with Mr. H. J. S. Smith; compare the article by Rosanes, "Ueber Systeme von Kegelschnitten" in *Math. Annalen*, vol. 6, page 264.

but is intersected by k in a self-polar quadrangle (kn) common to γ^2 and γ^2_1, which has either two or four real vertices. Finally, denote by δ^2 any one of the conics of the range, different from γ^2 and γ^2_1, which rests upon k, l, m, n (§§ 34, 35).

The theorem on the left is then proved by showing that upon any conic of the net of the third grade, for example, upon that one which passes through some three real points A, B, C (§ 34), there always rests, besides γ^2 and γ^2_1, also δ^2.

§ 40. The quadrangles (kl), (km), (lm), and (kn) are self-polar with respect to δ^2, since δ^2 rests upon each of the four conics k, l, m, n; consequently, δ^2 rests upon the four conics which are circumscribed to these four self-polar quadrangles and pass through any chosen point A. Three of these new conics are circumscribed to the self-polar quadrangle common to γ^2 and γ^2_1, determined by P and A (§ 38); the fourth does not pass through P if, as we may specify, A does not lie upon k; it therefore intersects the first three in three different self-polar quadrangles common to γ^2, γ^2_1, and δ^2. Among the four conics which pass through the point B and are circumscribed to these three self-polar quadrangles and to that determined by P and A, common to γ^2, γ^2_1, and δ^2, there are therefore at least two different ones supporting the curves γ^2, γ^2_1, and δ^2; these intersect in the common self-polar quadrangle of γ^2 and γ^2_1 of which A and B are two real vertices, and which is thus self-polar also with respect to δ^2; consequently, the conic passing through the point C and circumscribed to this latter self-polar quadrangle supports the curve δ^2. Thus the·theorem on the left of § 39 is proved.

§ 41. Two curves of the second class γ^2 and γ^2_1 lying in the same plane thus determine not only a net of conics of the third grade, but also a range of conics to which these two belong, whose curves rest upon all curves of the net.

Two curves of the second order k^2 and k^2_1 lying in the same plane determine not only a linear web of conics of the third grade, but also a sheaf of conics to which these two belong, upon whose curves all curves of the web rest.

The net contains an infinite number of conics which break up into pairs of straight lines; any straight line s of the plane together with that straight line s' which joins its poles with respect to γ^2 and γ^2_1 form one such line-pair. This degenerate conic supports all curves of the range of conics and its rays are conjugate with respect to each curve of the range (§ 32).

Hence,

The poles of a straight line *s* with respect to all curves of a range of conics lie upon a straight line *s'* ; these rays *s* and *s'* constitute a line-pair (degenerate curve of the second order) of the associate net of the third grade.	The polars of a point *S* with respect to all curves of a sheaf of conics pass through one point *S'* ; these points *S* and *S'* constitute a point-pair (degenerate curve of the second class) of the associate web of the third grade.

In particular, the centres of all conics of the range lie upon one straight line.

§ 42. If *s* is a common tangent of γ^2 and γ^2_1 it coincides with *s'* and is thus a two-fold line of the net. It is self-conjugate with respect to all curves of the range. Thus,

A straight line which touches any two conics of a range is a common tangent of all curves of the range and a two-fold line of the associate net of conics of the third grade. All the conics inscribed in a quadrilateral form a range of conics ; the associate net of the third grade contains all conics for which the quadrilateral is self-polar (§ 35).	A point through which two conics of a sheaf pass is a common point of all curves of the sheaf and a two-fold point of the associate web of conics of the third grade. All the conics circumscribed to a quadrangle form a sheaf of conics ; the associate web of the third grade contains all conics for which the quadrangle is self-polar.

§ 43. If a straight line *s* rotates about a point *A* its poles with respect to γ^2 and γ^2_1 describe two ranges of points *a* and a_1 projective to the sheaf *A*, and the line *S'* joining these poles describes in general a sheaf of rays of the second order. This contains all rays which are conjugate to the rays through *A* with respect to the range of conics (§ 41) ; it contains also the polars *a*, a_1 ... of *A* with respect to γ^2, γ^2_1, and every other curve of the range. These polars can be constructed by determining the poles of any two rays *g* and *h* through *A*, with respect to each of the curves of the range, and joining these pairs of poles. Since the lines joining these poles form a sheaf of the second order, we have (compare § 41)—

The poles of any two straight lines *g* and *h* with respect to the curves of a range of conics are homologous points of two projective ranges of points of the first order g_1 and h_1, and the polars of a	The polars of any two points with respect to the curves of a sheaf of conics are homologous rays of two projective sheaves of rays of the first order, and the poles of a straight line *a* lie in general upon a curve of

P

point A lie in general in a sheaf of the second order whose rays are conjugate to the rays through A, with respect to the range of conics.

the second order whose points are conjugate to the points of a, with respect to the sheaf of conics.

The centres of the curves of a sheaf of conics lie therefore in general upon a curve of the second order. In particular, the points of bisection of the six sides of a quadrangle and the points of intersection of the three pairs of opposite sides lie upon a conic.

In conformity with these theorems we may establish the following definitions :

Four conics of a range are said to be 'harmonic' if the poles of any straight line with respect to them are harmonic points.

Four conics of a sheaf are said to be 'harmonic' if the polars of any point with respect to them are four harmonic rays.

With these definitions it becomes possible to correlate ranges and sheaves of conics projectively to each other and to the elementary forms.

§ 44. Through a point A there pass in general two rays which are conjugate with respect to a range of conics (§ 43).

Hence,

The pairs of tangents which can be drawn from any point A to the curves of a range of conics form an involution. The two double rays of this involution are conjugate with respect to the range of conics, and each touches a conic of the range at A.

The pairs of points in which any straight line a intersects the curves of a sheaf of conics form an involution. The two double elements of this involution are conjugate with respect to the sheaf of conics, and in each of them a touches a conic of the sheaf.

§ 45. A range of conics is determined by any four conics of a net of the third grade upon which. it rests (§§ 34, 39), in particular by four line-pairs of the net. If we choose these line-pairs so that their four points of intersection lie upon one straight line l, it appears immediately that l is touched by a single curve of the range ; for a curve which touches l must also touch the four rays which are harmonically separated from l by the four chosen pairs of rays (§ 32). If two curves γ^2 and γ^2_1 of the range are given, the second tangent to this third curve, from any point A of l, can easily be drawn in accordance with the last section (§ 44).

From this it follows that—

An arbitrary straight line of the plane is touched by only one conic of a range ; but through an arbitrary point two conics of the range can pass (§ 44).	Through an arbitrary point of the plane there passes only one conic of a sheaf ; but an arbitrary straight line can be touched by two of these conics.

A range of conics therefore contains one, and only one, parabola unless all its curves are parabolas; the sheaf of conics contains two parabolas.

§ **46.** A straight line u whose poles with respect to γ^2 and γ^2_1 coincide at a point U (p. 110 Ex. 5) forms with each ray through U a degenerate conic of the net of the third grade and is the polar of U with respect to all curves of the range of conics (compare § 41). Hence (p 195, Ex. 14),

The curves of a range (or of a sheaf) of conics have in general a common self-polar triangle.

This is real and is easily constructed if any two of the conics have four real points or tangents in common (Art. 140) ; it has in any case at least one real side u and one real vertex U. If the conics γ^2 and γ^2_1 do not touch each other, U cannot lie upon its own polar ; and if the common self-polar triangle UVW of γ^2 and γ^2_1 is imaginary, the points V and W of u, which are conjugate with respect to the two conics, are imaginary. The pairs of points then in which γ^2 and γ^2_1 are intersected by u must in this case be real and must separate each other (Art. 248), and each of the conics γ^2 and γ^2_1 must enclose a portion of the other.

From this, and from what has previously been said, it easily follows that—

If the common self-polar triangle of a range or a sheaf of conics is imaginary, the conics themselves intersect by pairs in two and only two real points, and have two and only two real common tangents. These two tangents intersect upon the real side u of the self-polar triangle, and the two points lie in a straight line with the real vertex of the triangle.

For, otherwise, the intersection with u of the line joining the two points would be conjugate, with respect to the two conics, to the point U and also to a point of the joining line, and a second real vertex of the triangle would exist, hence the triangle itself would be real.

§ **47.** If a straight line s rotates about a point P of the straight

line u, its two poles with respect to γ^2 and γ^2_1 describe two projective ranges of points which have the point U self-corresponding,
and, consequently, the line joining these two poles rotates about a
point P'; since the two poles of PU lie upon u, P' must
also be a point of u. If now a conic k^2 which supports the two
conics γ^2 and γ^2_1 is passed through P, it must also pass through P';
for any two conjugate rays of P and P' intersect k^2 in a self-polar
quadrangle common to γ^2 and γ^2_1 (§ 29 b), and since at least
two sides of this quadrangle, different from u, pass through P,
the opposite sides conjugate to them (and hence also k^2) must
pass through P'. Since now u is cut in an involution by a sheaf
of conics which is circumscribed to any self-polar quadrangle of
the conics γ^2 and γ^2_1, it follows that—

Each real side u of the common self-polar triangle of a range of conics intersects the curves (and line-pairs) of the associate net of conics of the third grade, in pairs of points P, P', of an involution. The double elements of this involution u constitute a degenerate conic of the range, since they are conjugate with respect to all curves of the net.	From each real vertex U of the common self-polar triangle of a sheaf of conics there can be drawn as tangents to the curves (and point-pairs) of the associate web of conics, pairs of rays of an involution. The double rays of this involution U constitute a degenerate conic of the sheaf, since they are conjugate with respect to all curves of the web.

§ 48. The sheaves of rays P and P' (§ 47) are projectively
related if to each ray of the one is correlated the ray of the other
conjugate to it with respect to γ^2 and γ^2_1; they generate a curve
of the second order which is touched in P and P' by PU and $P'U$,
respectively. If the common self-polar triangle is real, this curve
passes through the double points of the two involutions v and w
which arise upon the other two sides of the triangle; and if the
double points of the involution u are imaginary, as when P and
P' are separated by v and w, one of the involutions v, w, has two
real double points lying upon this curve of the second order, and
the other, two imaginary double points. The rays through a real
double point O of u, v, or w are paired in involution in such a
way that any two conjugate rays of the involution are polar conjugates with respect to both conics γ^2 and γ^2_1, while O is the
point of intersection of two common tangents (real or imaginary)
of these conics, for instance, of the two double rays of the involution.

If the common self-polar triangle of γ^2 and γ^2_1 is imaginary, two real common tangents of γ^2 and γ^2_1 intersect in a point O of its real side u (§ 46), from which it follows that the involution u in this case has two real double points O.

Out of all this we have—

A range of conics contains at least one, but in general and at most three, real point-pairs ; these three point-pairs, whether real or imaginary, lie upon the three sides of the self-polar triangle of the range, and form the three pairs of opposite vertices of a real or imaginary quadrilateral to which the range of conics is inscribed.

A sheaf of conics contains at least one, but in general and at most three, line-pairs ; these three pairs of rays intersect in the three vertices of the self-polar triangle of the sheaf, and form the three pairs of opposite sides of a real or imaginary quadrangle to which the sheaf of conics is circumscribed.

§ 49. Besides the different limiting varieties whose curves touch or osculate in one point, we distinguish—

Three principal classes of ranges of conics, according as the curves of the range have four, or two, or no real common tangents. An imaginary self-polar triangle occurs only in the second of these classes (§ 46), and the three real point-pairs occur only in ranges of the first class. Two conics of a range of the first or the third class have either four or no real points of intersection ; two conics of a range of the second class always have two real points of intersection (§ 46).

If all curves of a range of conics degenerate into point-pairs, as is possible in a range of the second class, the range is a special one ; it is likewise special if it contains a two-fold point.

Accordingly, all conics which pass through the real or imaginary double points of an involution, or with respect to which any one point has a given polar, constitute a

Three principal classes of sheaves of conics, according as the curves of the sheaf have four, or two, or no real points of intersection. An imaginary self-polar triangle occurs only in the second of these classes, and the three real line-pairs occur only in sheaves of the first class. Two conics of a sheaf of the first or the third class have either four or no real common tangents ; two conics of a sheaf of the second class always have two real common tangents.

If all curves of a sheaf of conics degenerate into line-pairs, the sheaf is a special one ; it is likewise special if it contains a two-fold line.

Accordingly, all conics which touch the real or imaginary double rays of an involution, or with respect to which any one straight line has a given pole, constitute a special web of conics of the third

special net of conics of the third
grade ; a net is likewise special if
its conics all pass through one point.

grade ; a web of conics is likewise
special if all its conics have one
common tangent.

§ 50. The circles of a plane form a special net of the third grade ;
the associate range of conics consists of the pairs of points of the
involution in which the infinitely distant straight line cuts an orthogonal involution-sheaf. Confocal conics form a range of the third
class ; their two foci constitute the one real point-pair of the range,
and the associate net of conics of the third grade consists of all
equilateral hyperbolas with respect to which the two foci are conjugate. The circles which pass through two real points form a
sheaf of conics of the second class, and the circles orthogonal to
them form a sheaf of the third class. A special web of the third
grade is formed by the conics which have a given point F for
focus ; the conics of the associate sheaf degenerate into line-pairs
which intersect at right angles in F.

NETS AND WEBS OF CONICS OF THE SECOND GRADE.[*]

§ 51. Three curves of the second
class γ^2, γ^2_1, γ^2_2, which lie in one
plane but do not belong to a range
of conics, determine a net of conics
of the second grade (or briefly, a
net of conics) upon whose curves
they rest (§§ 34, 35).

Three curves of the second order
which lie in one plane but do not
belong to a sheaf of conics, determine a web of conics of the
second grade (or briefly, a web of
conics) whose curves rest upon
them.

All conics of the net which support, in addition to these three,
an arbitrarily chosen fourth curve of the second class, for example
a twofold point, form (§ 34) a sheaf of conics.

Hence (§ 45),

Through every point of the plane
there passes one sheaf of conics
belonging to the net ; through two
points chosen arbitrarily there
passes in general only one conic
of the net.

Every straight line of the plane
touches one range of conics belonging to the web ; two lines chosen
arbitrarily touch in general only
one conic of the web.

§ 52. Any two of the three curves γ^2, γ^2_1, γ^2_2, determine a net
of conics of the third grade containing the net of the second grade,

---

* Compare Schröter, *Die Theorie der Kegelschnitte*, 11. ed., pp. 500-535 (Leipzig,
1876).

and at the same time a range of conics whose curves rest upon all curves of the first net, hence also upon all curves of the second net (§41). The net of the second grade therefore supports the three ranges of conics determined by γ^2, γ^2_1, γ^2_2, and also every fourth range which contains any two curves of these first three ranges. All these ranges lie in the web of the second grade which is determined by three curves of the net chosen arbitrarily, for they rest upon these three curves; it may further be shown without difficulty that each curve of the web belongs to an infinite number of these ranges, and consequently rests upon every curve of the net of the second grade.

Hence the important theorem :

" With a net of conics there is always connected a web of conics "in such a way that each curve of the web rests upon every curve "of the net, and conversely. Three curves of the net or of "the web are sufficient in general to determine both these "manifolds of conics."

The net and the web of which we shall speak in the following sections are so connected that the web rests upon the net.

§53. A net of conics contains an infinite number of line-pairs which in general envelop a curve of the third class G^3, and intersect by twos in self-polar quadrangles common to γ^2, γ^2_1, γ^2_2, and the associate web of conics (§28). The rays of any such line-pair are conjugate with respect to the web, *i.e.*, with respect to all curves of the web (§32), and the poles of one of these rays all lie upon the other. Any two self-polar quadrangles of the web are always inscribed to one curve of the net; for that conic which is circumscribed to one of these quadrangles and passes through one vertex U of the other must belong to that sheaf of the net which passes through U, and therefore must pass through the remaining three vertices of the second quadrangle.

Similarly,

§54. A web of conics contains an infinite number of point-pairs ; these lie in general upon a curve of the third order C^3, and any two of them are pairs of opposite vertices of a self-polar quadrilateral of the associate net of conics. The points of such a pair are conjugate with respect to the net, *i.e.*, with respect to all curves of the net, and the polars of one of these points all pass through the other. Each point-pair of the web is harmonically separated by every line-pair belonging to the net. Any two self-polar quadrilaterals of the net are always circumscribed to one curve of the web. Of two conics, if one describes a sheaf the straight lines joining their points of intersection in general envelop a curve of the third class (§§ 52, 53); and, if one describes a range the points of intersection of their common tangents in general move upon a curve of the third order.

§ 55. A net of conics, together with its associate web of conics, is determined by three pairs of rays chosen arbitrarily which do not form the three pairs of opposite sides of a complete quadrangle (§ 52). If a sheaf of conics is circumscribed to the self-polar quadrangle of the web, in which two of these pairs of rays intersect, and we determine the intersections of the curves of this sheaf with the third pair of rays, we obtain all self-polar quadrangles of the web whose vertices lie upon these three pairs of rays (§ 53), and at the same time all line-pairs belonging to the net and the curve of the third class G^3 enveloped by them. Since now the third pair of rays may be any line-pair whatsoever of the net, and since its rays intersect the sheaf of conics in two point involutions (Art. 221),

"Any tangent to the curve of the third class G^3 is intersected "by the conics of the net in pairs of points of an involution. The "double points of this involution form a point-pair of the web and "lie upon the curve of the third order C^3."

For they are conjugate with respect to all curves of the net, since they are harmonically separated by three of these curves chosen arbitrarily.

§ 56. In the same way, a web of conics together with its associate net is determined by three pairs of points chosen arbitrarily which do not form the three pairs of opposite vertices of a complete quadrilateral, and from these all remaining point-pairs of the web, and the curve of the third order C^3 upon which they lie, can be derived.

The pairs of tangents which can be drawn from any point P of the curve of the third order C^3 to the conics of the web form an involution. The double rays of this involution constitute a line-pair of the net and touch the curve of the third class G^3.

From a point P chosen arbitrarily upon the curve C^3 the point-pairs of the web are projected by an involution sheaf.	By an arbitrary tangent to the curve G^3 the line-pairs belonging to the net are cut in the points of an involution.

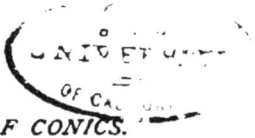

The points of intersection of all line-pairs of the net lie upon C^3; on the other hand, G^3 is enveloped by the joining lines of point-pairs of the web (compare § 61).

§ 57. The points of C^3 are conjugate, two and two, with respect to the net, and the tangents to G^3 are conjugate, two and two, with respect to the web (§§ 53, 54).

We shall now define as follows:

Three points of C^3, whose conjugates lie upon one straight line, shall be called a 'point-triple' of C^3.

Three tangents to G^3, whose conjugates pass through one point, shall be called a 'tangent-triple' of G^3.

Since two point-pairs of the web always form two pairs of opposite vertices of a self-polar quadrilateral of the net determined by them (§ 54), we have immediately:

The three points of any point-triple of C^3, together with their conjugate points, form the three pairs of opposite vertices of a self-polar quadrilateral of the net; every self-polar quadrilateral of the net contains four point-triples of C^3. Any two points of C^3 whose joining line passes through a given point P_1 of C^3 form with the conjugate point P of P_1 a point-triple of C^3.

The three tangents of any tangent-triple of G^3, together with their conjugate tangents, form three pairs of opposite sides of a self-polar quadrangle of the web; every self-polar quadrangle of the web contains four tangent-triples of G^3. Any two tangents of G^3 which intersect upon a given tangent t_1 of G^3 form with the conjugate tangent to t_1 a tangent-triple of G^3.

§ 58. The two triangles formed by any two point-triples of C^3 are always circumscribed to a conic of the web (§ 54) since they lie in two self-polar quadrilaterals of the net (§ 57).

But from this it follows (see p. 81, Ex. 10) that—

Any two point-triples of C^3 are inscribed in a curve of the second order.

If then a conic is circumscribed to a point-triple of C^3 and passes through two other points P and Q of the curve, it is also circumscribed to the point-triple of C^3 determined by P and Q.

Any two tangent-triples of G^3 are circumscribed to a curve of the second class.

If then a conic is inscribed in a tangent-triple of G^3 and touches two other tangents p and q of G^3, it is also inscribed in the tangent-triple of G^3 determined by p and q.

§ 59. The second triple in the theorem on the left of the last section coincides with the first if P and Q approach indefinitely near to two points of the first triple. Hence,

To every point-triple of C^3 a conic can be circumscribed which touches the curve C^3 in the three points of the triple.	To every tangent-triple of G^3 a conic can be inscribed which touches the curve G^3 at the points of contact of these tangents.

From the theorems of §57 we further obtain corollaries as follows :

The tangents at P and P_1, two conjugate points of C^3, intersect in a point Q of C^3 which with P and P_1 constitute a point-triple of C^3.	The points of contact of two conjugate tangents t and t_1 of G^3 lie upon that tangent of G^3 which together with t and t_1 constitute a tangent-triple.

§ 60. It follows from § 59 and the concluding statements of § 57 that—   .

Any two points of C^3 whose joining line passes through a given point P_1 of C^3 lie upon a conic of that sheaf which is circumscribed to the quadrangle formed by an arbitrary point-triple of C^3 and P the conjugate point of P_1.	Any two tangents to G^3 which intersect upon a given tangent t_1 of G^3 touch a conic of that range which is inscribed in the quadrilateral formed by an arbitrary tangent-triple of G^3 and the tangent t conjugate to t_1.

In the theorem on the left, the sheaf of rays P_1 and the sheaf of conics are so related projectively to each other that they generate the curve C^3, and similarly, in the theorem on the right, the range of points t_1 and the range of conics generate the curve G^3.   But it is not my intention to prove the truth of this statement at this point.

§ 61. The sides of the self-polar triangle common to two conics k^2 and k^2_1 of a net belong to and determine a self-polar quadrilateral of any third conic k^2_2 of the net (§ 24).   This is then a self-polar quadrilateral of k^2, k^2_1, and k^2_2, and hence of the net (compare § 57).

The vertices of a self-polar triangle common to any two conics of the net always form a point-triple of the curve C^3. In particular, the three pairs of opposite sides of any self-polar quadrangle of the web intersect in a point-triple of C^3.	The sides of a self-polar triangle common to any two conics of the web always form a tangent-triple of the curve G^3. In particular, the three pairs of opposite vertices of any self-polar quadrilateral of the net lie upon a tangent-triple of G^3.

The first half of this theorem is converse to itself. It is easily proved further that every point whose polars with respect to two curves of the net coincide lies upon C^3, and that every straight line whose poles with respect to two curves of the web coincide is tangent to the curve G^3.

§ **62.** The straight lines which intersect any three given conics of a plane in three pairs of points in involution envelop a curve of the third class G^3 (§ 55); this curve also touches the lines joining the points of intersection of the conics, two and two (§ 53).

The points from which tangents to any three given conics of a plane form three pairs of rays in involution lie upon a curve of the third order C^3 (§ 56); this curve passes through the points of intersection of tangents common to the conics, two and two (§ 54).

§ **63.** Since two conjugate points Q and Q_1 of the curve C^3 are harmonically separated by all pairs of conjugate tangents to the curve G^3 (§ 54), and since to the point Q_1 of C^3, which lies in a straight line with two conjugate points P and P_1, is conjugate the point of intersection Q of the tangents at P and P_1 (§§ 57, 59),

Any tangent to C^3 and the three straight lines which can be drawn from its point of contact P to touch G^3 form a harmonic sheaf of rays;

Any point of G^3 and the three points in which the tangent at this point intersects C^3 form a harmonic range of points;

and the tangent to C^3 is harmonically separated from PP_1 by two conjugate tangents to the curve G^3. If P_1 is a point of inflexion of C^3, Q coincides with P_1, and PP_1 is tangent to C^3 at P, and conversely. But in accordance with the theorem on the right, the point in which G^3 is met by the tangent PP_1 coincides with P. Hence the theorem:

"The curves C^3 and G^3 touch each other at all common points; "the tangent at any common point P intersects the curve C^3 at "one of its points of inflexion P_1 and is intersected at P by an "inflexional tangent of the curve G^3."

§ **64.** In the web of conics there is contained one range of parabolas (§ 51); the foci of the parabolas in this range lie upon a circle, while their directrices intersect in a point K (compare p. 179 Ex. 3). The net contains in general one circle (centre at K), one sheaf of equilateral hyperbolas, and an infinite number of parabolas of which not more than two pass through any .one point (§§ 45, 51).

§ **65.** Nets of conics may be divided into four main classes,

between which a number of varieties of special nets form a
transition. We distinguish these main classes by the aid of the
involutions in which the net is cut by the straight lines of its
degenerate conics (§ 55). If these involutions are in part elliptic
and in part hyperbolic, *i.e.*, if some of the involutions have
imaginary and some real double points, their bases envelop two
different branches of the curve G^3; for the net is intersected by
successive tangents to this curve in involutions of the same
kind since the double elements lie upon the curve C^3. The two
branches of G^3 have a common tangent t_1 only when the two
double points of the involution upon t_1 coincide; but since these
double points are conjugate with respect to all conics of the
net, the conics in the assumed case all pass through one (self-
conjugate) point and the net is of a special character.

§ 66. A non-special net of conics is either cut by both rays
of all its degenerate conics in hyperbolic involutions, or by the
rays of some of them in elliptic involutions, or, finally, by one
ray of each pair in a hyperbolic involution and by the other
in an elliptic involution. I shall call the net in the first case
hyperbolic, in the second elliptic, and in the third dual. A net is
hyperbolic if it contains a sheaf of conics of the third class (§ 49),
or, in general, two conic sections which have no real points in
common; for, a sheaf of the third class is intersected by the lines
of the plane in hyperbolic involutions. The conics of an elliptic
or dual net intersect, two and two, in at least two real points.

§ 67. If the net is given by three of its degenerate conics, we
can select two of these as two pairs of opposite sides of a quadrangle contained in the net. By the third pair, and in general
by any conic of the net, either one vertex or no vertex of this
quadrangle is separated from the remaining three, according as
the net is dual or hyperbolic. On the other hand, if the net
is elliptic, either two or no vertices of the quadrangle are
separated from the others by the third line-pair, and in general
by the conics of the net; certainly two, if any pair of opposite
sides of the quadrangle intersect the net in elliptic involutions.

§ 68. A dual net is cut by the two rays of any of its line-pairs
in involutions of different kinds, while, on the contrary, a hyperbolic or elliptic net is cut in involutions of the same kind.
Every quadrangle contained in a dual net has one vertex which
is separated from the remaining three by the line-pairs and the

conics of the net; the three sides of the quadrangle which pass through this vertex intersect the net in elliptic involutions, the remaining three sides in hyperbolic involutions.

In an elliptic net there is an infinite number of quadrangles, of which two pairs of opposite sides intersect the net in elliptic involutions, and the remaining pair in hyperbolic involutions. The two pairs of vertices of such a quadrangle which lie upon the latter two sides are separated by the conics of the net.

In a hyperbolic net there is no quadrangle whose vertices are separated by any conic of the net.

§ 69. A web of conics is projected from the points P of its point-pairs by involution sheaves, *i.e.* through each of these points P there pass pairs of tangents to the conics which form an involution. In particular, the point-pairs of the web are projected from the points P by pairs of rays of these involutions. If some of the involutions are elliptic and others hyperbolic, their centres P lie upon different branches of the curve of the third order C^3, and only in special cases are these two branches connected in a double point (compare § 65). If all of these involutions are hyperbolic, we call the web hyperbolic; if they are in pairs partly elliptic and partly hyperbolic, the web is said to be elliptic ; if, finally, the web is projected from one point of any point-pair, and hence of every point-pair, by an elliptic involution and from the other by a hyperbolic involution, it is called a dual web.

§ 70. I shall denote the dual net of conics as a 'net of the first kind.' Its curve of the third class G^3 consists of an odd and an even branch. The net is cut by the tangents of the even branch in hyperbolic involutions, and by those of the odd branch in elliptic involutions; from a vertex of any quadrangle contained in the net there can be drawn one or three tangents to the odd branch, and to the even branch either two or none (§ 68). Two conjugate tangents to G^3 never touch the same branch of the curve ; if they simultaneously glide along G^3 their point of intersection describes the curve C^3, which is 'unicursal,' *i.e.*, it consists of only one real branch. In every real point P of C^3 two real conjugate tangents of G^3 intersect, and the web resting upon the net is accordingly projected from the points P of its point-pairs by hyperbolic involutions of which the pairs of real conjugate tangents to G^3 are the double rays. The dual net is therefore the support of a hyperbolic web, whose point-pairs lie upon a unicursal curve of the third order C^3. The tangents to the even branch alone of G^3 connect two real conjugate points of C^3.

§ 71. The elliptic net of conics is designated a 'net of the second kind.' Its curve of the third class G^3 consists of an odd and an even branch; the net is intersected by the tangents of the even branch in elliptic involutions and by those of the odd branch in hyperbolic involutions. Either two opposite sides only, or else all sides of a quadrangle contained in the net, intersect the net in hyperbolic involutions, so that through each vertex of the quadrangle there pass either one or three bases of hyperbolic involutions, and either two or no bases of elliptic involutions. Of two conjugate tangents to G^3, either both touch the odd branch or both the even branch of the curve; in the first case, their point of intersection lies upon the even branch of the curve C^3, in the second case, upon the odd branch of C^3, with which they now have no other real points in common. The curve of the third order C^3 thus consists, for the net of the second kind, of two different branches, and in each real point P of the curve two real conjugate tangents of G^3 intersect. It follows from this (compare § 70) that the elliptic net is the support of a hyperbolic web. The tangents to the odd branch alone of G^3 connect two real conjugate points of C^3, and it is evident that one of these two points lies upon the odd branch and the other upon the even branch of C^3.

§ 72. A hyperbolic net of conics shall be called a net of the third or fourth kind, according as the web resting upon it is dual or elliptic.

The net of the third kind is therefore the reciprocal of a web which rests upon a net of the first kind. Its line-pairs envelop a unicursal curve G^3, while C^3 consists of two different branches. Every real tangent of G^3 joins two real conjugate points of C^3 which are distributed upon both branches of C^3; through a real point of C^3, on the other hand, there pass two real or two imaginary conjugate tangents of G^3, according as the point lies upon the even or upon the odd branch of C^3. The net contains sheaves of conics of the first, second, and third classes; the web resting upon it contains ranges of conics of the first and second classes, but none of the third class.

§ 73. The net of the fourth kind is the reciprocal of a web which rests upon a net of the second kind. The associate curves C^3 and G^3 consist therefore of two different branches each (§ 71).

Every real tangent to G^3 joins two real conjugate points of C^3; these lie upon the odd or the even branch of C^3 according as the tangent touches the even or the odd branch of G^3. Through a real point of C^3 there pass two real or two imaginary conjugate tangents to G^3, according as the point lies upon the odd or upon the even branch of C^3. The net contains sheaves of conics of all three classes; the web resting upon it contains ranges of only the first and second classes.

§ 74. No others than the four principal kinds of nets of conics mentioned above can exist, since a hyperbolic net is never the support of a hyperbolic web. In a hyperbolic net, for instance, there are always imaginary line-pairs with real intersections P; but from such a point P the associate web is projected by an elliptic involution sheaf of which this imaginary line-pair forms the double rays, and the web accordingly is not hyperbolic.

If a non-special net supports a web, then of these two manifolds of conics, the one is always hyperbolic and the other either elliptic or dual (§ 70 to § 73). Both or at least one of the two associate curves G^3 and C^3 consists of two branches.

§ 75. The web is a special one if it contains a two-fold point Z; such a point becomes a double point of C^3 and through it pass all the conics of the associate net, which is likewise a special one, and the curve G^3 breaks up into the sheaf of rays Z and that curve of the second class G^2 whose tangents are conjugate to the rays of Z with respect to any two curves of the web. The sheaf of rays Z and the sheaf of tangents to G^2 are projectively related and generate the curve C^3 (§ 56).

The web and the net are again special if the net contains a two-fold line z; this line is a double tangent to G^3 and touches all the conics of the web; the curve C^3, in this case, degenerates into this twofold line and a curve of the second order.

§ 76. The web is also special if it contains all pairs of points of an involution u. In this case the conics of the net all pass through the two double points of u, the curve G^3 degenerates into these two double points and the pole U of u with respect to any curve γ^2 of the web, and C^3 degenerates into the straight line u and a curve of the second order (compare p. 162, Ex. 14). The straight line u is conjugate, with respect to the web, to all rays of the sheaf U, and through U there passes a common chord of every pair of conics of the net. If, in particular, the involution u is the

section of an orthogonal involution sheaf made by the infinitely
distant line, then the net consists wholly of circles (compare § 50).

§ 77. The circles which support a given curve of the second class
γ^2, *i.e.*, which can be circumscribed to its self-polar triangles and
quadrangles, form a special net of conics. The chords which are
common to these circles two and two intersect in the centre U of γ^2;
they are parallel if γ^2 is a parabola. If γ^2 is an ellipse or hyperbola,
and hence the product of the segments of the secants drawn to the
circles from the centre U, constant, the circles have equal 'powers'
at U and there exists a circle (imaginary, however, for obtuse-
angled hyperbolas) concentric with γ^2 which intersects all circles of
the net orthogonally. If on the other hand γ^2 is a parabola, the
centres of all circles of the net lie upon one straight line. Each
point of this straight line, or of the orthogonal circle concentric with
γ^2, may be considered an indefinitely small circle of the net, and
from this we conclude that the tangents from any one of these
points to γ^2 are at right angles to each other.

The theorems of this section were discovered by Faure.

§ 78. From the reciprocal to § 76 we have, among other things,
the following :

The conics which have a given point F for focus and rest upon
a curve of the second order k^2 form a special web. The two
real tangents which any two of them have in common always
intersect upon the polar f of F with respect to k^2. The curve G^3
breaks up into F and a curve of the second class; C^3, on the
other hand, breaks up into f and the two imaginary double rays
of the orthogonal involution sheaf F. Two opposite sides of each
self-polar quadrangle of the web intersect at right angles in F;
the points of intersection of the remaining two pairs of opposite
sides lie upon f. In general, f is the polar of F with respect to
all conics of the net upon which the web rests.   '

§ 79. The conics which can be circumscribed to a triangle ABC
form a highly specialized net ; of the curves of the associate web,
ABC is a common self-polar triangle ; the curves C^3 and G^3 reduce,
respectively, to the three sides and the three vertices of the triangle.
The conics of which ABC is a self-polar triangle form not only a
highly specialized web, but a highly specialized net as well; upon this
net rests the web which can be inscribed in the triangle.

INDEX.

(The numbers refer to pages.)

Q

GLASGOW: PRINTED AT THE UNIVERSITY PRESS BY ROBERT MACLEHOSE AND CO.

www.ingramcontent.com/pod-product-compliance
Lightning Source LLC
Chambersburg PA
CBHW021519210326
41599CB00012B/1308